Finding the
# UFO Crash
at San Augustin
Isotopic Metal Analysis
Not of This World

by
Arthur H. Campbell

Medford, Oregon

# Finding the UFO Crash at San Augustin: Isotopic Metal Analysis – Not of This World

© 2014 Arthur H. Campbell

Expanded and revised edition based on The UFO Crash at San Augustin, © 2002 by Arthur H. Campbell

All Rights reserved, including the rights of reproduction in any form without written or contractual consent from author. These rights are reserved under the Pan American International Copyright Conventions and include information storage and retrieval systems now known or here after invented. This also includes copying by any means electronic, mechanical, photo copying and/or audio recording.

Finding the UFO Crash at San Augustin

ISBN number 978-1491221945

Printed by Create Space, an Amazon.com company.
Available from other retail outlets including CD and E-book formats.

Campbell, Arthur H.
    Finding the UFO Crash at San Augustin: Isotopic Metal Analysis – Not of This World / Arthur Campbell

    Photo Credits:
UFO Magazine, MBI Publishing, Alice Knight, Beth Danley, Verne Maltais, The Smithsonian, Adams State College, Indiana University, Walter Reed Medical Center,

## Dedication

To my wife of 50 plus years, Mary Lou, who has endured the dust and critters of the Plains of San Augustin (real and imagined), a moody husband and time behind a cranky typewriter and later a cantankerous computer. A sincere thanks for her support and tacit permission to go off to New Mexico for "one more trip." She was with me in 1958 when both our marriage and UFO investigations began, and with me over 50 years later when we celebrated our 50th anniversary and beyond. Thank you, Mary Lou.

"In order for any material to be considered a genuine extraterrestrial artifact, three main characteristics must be satisfied. First, the testing must provide conclusive results that the elemental composition of the material is of extraterrestrial origin and could not have come from this world. Secondly, it must have uniform structure. And third, the laboratory test must prove the material was manufactured and not naturally formed. That is, it must not be a meteorite or meteorite fragment."

— *Dr. Russell Vernon Clark*
*Chemist from the University of California San Diego*

## Table of Contents

**Acknowledgments** ..................................................................................................................vii
**Introduction** ..........................................................................................................................ix
**Chapter 1. The Plains of San Augustin** ...............................................................................1
    The Birth of an Arroyo ........................................................................................................3
**Chapter 2. The Site** ...............................................................................................................7
    Rattlesnakes .........................................................................................................................7
    Hantavirus ...........................................................................................................................8
    The Soil Analysis ................................................................................................................9
    Chuck Wade .......................................................................................................................11
**Chapter 3. A Cross Marked the Spot — The First Dig** ....................................................13
    Strange Materials Found ....................................................................................................15
**Chapter 4. The Foil Shards "May Not Have Originated on Earth"** ..............................17
    Overview ............................................................................................................................17
    Laboratory Equipment .......................................................................................................19
    The Colbern Report — Background .................................................................................22
    The Colbern Report — The Samples ................................................................................23
    The Colbern Report — Discussion ...................................................................................24
    The Colbern Report — Conclusions .................................................................................25
    Cutting Edge Equipment — Timelines .............................................................................26
**Chapter 5. Researchers, Gerald & the Motherlode** ........................................................27
    Finding the Motherlode .....................................................................................................29
    Nature's Smallest ...............................................................................................................31
    The Honeycomb ................................................................................................................32
    No Beer Can ......................................................................................................................32
    Skipping In ........................................................................................................................33
    Military Minds ..................................................................................................................34
    The Honeycomb Sandwich ...............................................................................................35
    The Shikoku Island Honeycomb ......................................................................................36
    Evidence of Repairs ..........................................................................................................37
    Other Concerns .................................................................................................................37
    Metal Sample Analysis by Steve Colbern .........................................................................38
**Chapter 6. The I-Beam, Second Honeycomb & Propulsion** ...........................................39
    65 Years Later ...................................................................................................................39
    I- and H-Beams .................................................................................................................40
    The Triangle/Circle Symbol .............................................................................................40
    Sky Path & Eclipse Surprise .............................................................................................41
    Rendelsham Forest Connection ........................................................................................41
    The Second Dig Crew .......................................................................................................42
    Site History, Early Finds ...................................................................................................42
    Loss of Power ....................................................................................................................43
    Second Honeycomb ..........................................................................................................43
    San Augustin Honeycomb & Significance — by Steve Colbern ......................................44
    Rare Earth Metals .............................................................................................................45
    Craft Propulsion ................................................................................................................46
    Brown Dot Propulsion Theory .........................................................................................46
    Summary of I-Beam & Brown Dots .................................................................................47
**Chapter 7. Other Finds — "That Ain't No Part of No Cow"** ........................................49
    What Is That? ....................................................................................................................51
    Other Finds .......................................................................................................................51
    Little Shoe Sole .................................................................................................................52
    The Fabric Scraps .............................................................................................................52
    Cup of Coffee and a Zinc Wafer .......................................................................................53

## Chapter 8. The Grady Barnetts in 1947 ........................................................................... 55
    The Early Years .................................................................................................................. 57
    Engineering ......................................................................................................................... 59
## Chapter 9. Barney's Work, Travels & Discovery .............................................................. 63
    Barney's Work and Travels ................................................................................................ 64
    July 1947, the Discovery ................................................................................................... 66
    The Roswell Events and Barney's Travels ........................................................................ 69
    Confirmation ...................................................................................................................... 71
    Another Confirmation? ...................................................................................................... 73
## Chapter 10. Letter from Harvard, the Archaeologists & the Meteor .............................. 75
    Written Off ......................................................................................................................... 75
    The Conflict ....................................................................................................................... 76
    A Meteor? .......................................................................................................................... 77
## Chapter 11. Bat Cave and Beyond ...................................................................................... 81
    Other Archaeologists ......................................................................................................... 84
    Survey Results at Bat Cave ................................................................................................ 87
    UFO Intrigue ...................................................................................................................... 88
## Chapter 12. Artifact & Contents — Closer Exam .............................................................. 91
    First Tests ........................................................................................................................... 91
    In the Lab ........................................................................................................................... 92
    Observations — More Tests .............................................................................................. 93
    Body Fluid & Starch .......................................................................................................... 95
    The Gold Wire and the Redheaded Co-ed ......................................................................... 96
    A VAD & Tupperware ....................................................................................................... 98
## Chapter 13. The Shoe Sole, Wax Pieces & Wafer ............................................................ 101
    Lab Results ....................................................................................................................... 101
    Shoe Companies ............................................................................................................... 102
    UFO Literature ................................................................................................................. 105
    Human Foot Development ............................................................................................... 106
    ET Bodies ......................................................................................................................... 106
    The Wafer ........................................................................................................................ 107
    The Wax Material ............................................................................................................ 108
    Review ............................................................................................................................. 109
## Chapter 14. The Impact — Out of Sight, Out of Mind .................................................... 111
    The Gap ........................................................................................................................... 111
    The HDPE Artifact .......................................................................................................... 113
    Colonel Corso .................................................................................................................. 114
    The Gray Alien Clone ...................................................................................................... 115
    Natural Material ............................................................................................................... 116
    The Shoe Sole .................................................................................................................. 116
    Summary .......................................................................................................................... 117
## Chapter 15. Plane Crashes, Chickens & Missiles ............................................................ 119
    Missiles ............................................................................................................................ 121
    Where Are They Now? .................................................................................................... 123
## Chapter 16. Bringing It All Together ............................................................................... 125
    The Land Transfer ........................................................................................................... 125
    Security ............................................................................................................................ 126
    Transporting the Craft ...................................................................................................... 128
    Corroborations ................................................................................................................. 129
    The Disturbed Rock Cairn/Grave .................................................................................... 131
    The Last Moments ........................................................................................................... 133
    Writing Off Barney Barnett .............................................................................................. 134
    Reflections ....................................................................................................................... 135
## Index ..................................................................................................................................... 137

# ACKNOWLEDGMENTS

I am grateful for the many groups, organizations, and individuals who contributed to this documentary. There may be an omission of certain people, as well as specific places. Controversy is an area of avoidance by Western people, and in order not to embarrass some individuals and families who helped with this project, names of many participants will be omitted. The very nature of information gathering requires building a trust between the writer and the source. I will, in general, thank many through their parent organizations. In some cases, more than one person in an organization was helpful. As I did not mention UFO research to most of those I dealt with, I feel it only fair and reasonable not to identify them with this project. I do earnestly hope that the merits of this work will be judged by the scientific evidence and documentation presented, rather than the continuous telling of the same old UFO bedtime story with its time-worn pages and familiar heroes.

Much of the historical background research was done at various libraries. I wish to thank the New Mexico State Library at Santa Fe, New Mexico, for its cooperation with the inter-library loan system. Through them I ordered and viewed hundreds of newspapers on microfilm. Also thanks to the Socorro, New Mexico Public Library and the Library of the National Radio Astronomy Observatory in Socorro, New Mexico and the Health Science Library at the University of Virginia, Charlottesville, Virginia. Best wishes and "hearty thanks" to Randy Sowell and Sam Rushay, archivists at the Harry S. Truman Library in Independence, Missouri.

Museums were another important source of information. Thanks to the New Mexico Farm and Ranch Museum in Las Cruces, New Mexico; the Cimarron Heritage Center, Boise City, Oklahoma; the Armstrong County Museum, Claude, Texas; the Humboldt County Historical Association, Humboldt, Iowa; War Eagles Air Museum, Santa Teresa, New Mexico; Square House Museum, Panhandle, Texas; The New Mexico Museum of Indian Arts and the Alabama Museum of Natural History. Many thanks to the Smithsonian Natural History Museum in Washington, D.C. for opening its vaults and letting me view the first artificial human heart.

Appreciation is extended to the Texas Department of Health, Austin, Texas; Arizona Department of Health Services, Phoenix; State Aviation Department, New Mexico Aviation Division, Santa Fe. Other New Mexico contributors were: the State Highway and Transportation Departments at Roswell and Albuquerque, the Commission of Public Records, State Records and Archives Center in Santa Fe, and the Socorro Electrical Co-op. Thanks is also given to the Navopache Electrical Coop, Lakeside, Arizona. And a special thanks to Carol Baird of Telecommunications, Denver, Colorado, for vintage 1947 New Mexico telephone book information.

For New Mexico and federal land records, thanks goes to the Socorro County Surveyor's Office in Socorro, the Catron County Surveyor's Office in Reserve, and the Chaves County Tax Assessor's Office in Roswell. Also of assistance was the Bureau of Land Management in Las Cruces and the old military and civil records branch, Textual Reference Branch, Division of the National Archives, Washington, D.C. The author greatly appreciates the assistance of the US Department of Agriculture, archives section in Albuquerque.

Among the schools of higher learning that I wish to thank are: The School of American Research, Santa Fe, New Mexico; and the University of New Mexico Department of Archaeology and the Department of Ecology, as well as the University of New Mexico Maxwell Museum at Albuquerque.

The aluminum foil companies were very helpful; they include: Alufoil Products Company, Inc., Hauppauge, New York; Winter-Wolff International Inc., Jericho, New York; Alcan Chemicals, Division of Alcan Aluminum, Cleveland, Ohio; Alcoa Aluminum, Pittsburgh, Pennsylvania; and A.T. Oster Foils Inc., Alliance, Ohio. A special thanks goes to Paul Mara, Aluminum Association in Washington, D.C., for assistance in researching the metal foil.

Thanks goes to the following companies: Amco Plastic Materials Inc., Farmingdale, New York; Arco Chemical Company, Newtown Square, Pennsylvania; Dow Plastics, Midland, Michigan; Phillips Chemical Company, Bartlesville, Oklahoma; Plastic Central-Materials, Kalamazoo, Michigan; Nicholet Instrument Corporation, Madison, Wisconsin; Interstate Plastic, Post Falls, Idaho; and Chemical Manufacturing Association, Rosslyn, Virginia. And a special thanks goes to the National Shoe Retailers' Association in New York City; the SATRA Footwear Technology Center, North Hamptonshire, United Kingdom; the Western Shoe Association; the Footwear Retailers' Association; and the Eastman Kodak Company of Rochester, New York.

The following government offices in Socorro, New Mexico, were of valuable assistance: the Bureau of Land Management, the Geological Information Center and the New Mexico Bureau of Mines and Mineral Resources. Also of assistance were the Office of the State Geologist, New Mexico Institute of Mining and Technology, and personnel of the U.S. Department of Agriculture. The assistance of these organizations was greatly appreciated. The Natural Resources Conservation Service (NRCS), located in Albuquerque, Clayton, Roy, Capitan, Carrizozo and Roswell, New Mexico. Thanks to the Cibola National Forest, Magdalena Ranger District, Magdalena, New Mexico. I also wish to thank the U.S. Department of the Interior, Geological Survey in Tucson, Arizona and Denver, Colorado. Military offices and personnel were of great assistance. I would like to thank the Military History Institute, Research and Reference Branch, Army War College, Carlisle Barracks, Carlisle, Pennsylvania; the base library, Fort Leonard Wood, Missouri; the United States Air Force Museum, Wright-Patterson Air Force Base, Dayton Ohio; Center for Military History, Historical Support Branch, Fort McNair, Washington, D.C.; Major General Joseph L. Dickman, Air Force History Office, Bolling Air Force Base, Washington, D.C.; Lynn Gamma, Archivist, Air Force Historical Research Agency, Maxwell Air Force Base, Alabama; the Department of the Army Center for Military History, Washington, D.C.

I wish to send hearty thanks to a number of landowners on and near The Plains of San Augustin. I have agreed not to identify the various sites or property owners. On one occasion or another, I have worked with five separate owners of one piece of property. I especially appreciated a timely cup of coffee and occasionally something stronger, and also a hearty meal now and then. I greatly appreciate the trust and confidence of those who had little to gain by my presence on their land and much to lose. Later on, as more investigative team members came to the site, it became much more difficult to keep the location quiet and secure. I am especially grateful for information gleaned from the descendants of James A. Hubbell. In 1947, the Hubbell Sheep and Cattle Company owned the Plains of San Augustin's Y Ranch as well as Bat Cave. Archaeologists working at the cave in 1947 may have been witnesses to a Plains UFO crash. We wish to thank two very nice ladies from The Plains area. They are Trudy Ake, whose husband Marvin knew and respected Barney Barnett. The other lady, Mrs. J.F. (Beth) Danley, also knew the Barnetts and shared personal insights about her husband Fleck's role in the Plains of San Augustin crash aftermath.

Five witnesses corroborated Barney Barnett's story about finding a crashed UFO on The Plains. Their stories told once again with personal data are indispensable to this work. Among them are: Alice Knight, Barney's niece, and Mr. and Mrs. W.L. (Vern) Maltais, a long-time personal friend of the Barnetts. Another independent key witness who verified The Plains crash was Colonel William Leed. I also wish to thank two nice people who were the neighbors of the Barnetts, Harold and Martha Baca.

Once in a while during mundane research one gets a breath of fresh air; in this case it was Mrs. Leonard Stringfield of Cincinnati, Ohio. Mrs. Stringfield's positive manner and generous permission to use her late husband's research were greatly appreciated. Another friendly and welcome individual was John A. Price, who ran the now-closed UFO Enigma Museum in Roswell, New Mexico. Mr. Price's efforts to bring UFO information to the public will be sorely missed. The author is especially grateful to Dr. Wesley R. Hurt, Jr., for sharing some recollections, insights, and material from his undergraduate days at the University of New Mexico.

The UFO Field had its supporters also. A sincere thanks goes to Bill Birnes of UFO Magazine and a co author with Philip Corso of the best selling book The Day After Roswell. Later in the project, we could not have gone forward, without the help and encouragement of Nuclear Physicist Stanton T. Friedman, to whom the UFO community owes much. It was at a lecture in Bemidji, MN where Vern and Jean Maltis approached Mr. Friedman and told him about Barney Barnett's Plains crash experience. Stanton was also the first to speak to William Leed, who interviewed Barnett and confirmed his crash experience from Pentagon sources. A sincere thanks goes to the UFO information retrieval center Inc. of Riderwood Maryland.

We wish to thank Harvard University's Peabody Museum of Cambridge, MA. When an early draft of this book was finished in 2005, permission was given to include crucial material from the files of archaeologist Herbert W. Dick. Dick and his party's denial of their presence on the Plains the first week of July 1947, is refuted by Dick himself in a letter dated Dec. 14, 1947 (See p. 108 and Appendix L). Harvard University is the first college library to acquire copies of Finding the UFO Crash at San Augustin.

Any complete work such as this has many behind the scenes individuals who were indispensable. They include Kristi and Shannon from the local camera store, the knowledgeable helpers at the copy center. Other dedicated people were Sean Seville,, my friend Jobie Grether, and his artists for their splendid work A sincere thanks goes to Dorothy and Martin Zimmerman, Louise Watson, Paul Bartlett, and Mildred Icy in Gravet, Arkansas. We wish to give a hearty thanks to Randy Haragan who for years operated a very nice UFO store and website. His early encouragement was significant.

Once in awhile someone outstanding appears on the UFO scene. Tony Beragalia is known for his blog, UFO Iconoclast(s). Tony is an excellent researcher and very thorough. His readership is thoughtful and intelligent. His article in 2010 on the Plains events and personalities, including Barney. Barney was extremely well done. Reader comments encouraged him to "keep chipping away." Another articulate reader observed he was an expert at "super imposition of theoretical overlays." Tony is the executive employment field and is a resident of Florida.

The analysis of the materials found at the crash site were supported and encouraged by over a dozen laboratories and key individuals. We wish to especially thank Dr. Roger Leir, podiatrist and UFO implant researcher for his wisdom and counsel. We especially appreciate the leadership of the Open Minds organization for their help and assistance in getting the metal shards initially analyzed. Without good, competent scientists, the extraterrestrial nature of our found materials could not have proceeded without the work of chemist and materials scientist Steve Colbern. He did a great job of making the complex scientific research in this book palatable to the reader.

There have been many competent computer people involved in this work. Many have been mentioned above. One gal deserves special recognition, who brought her expertise and portable laptop to finishing the last stages of this book. We are extremely grateful for Vanessa Brook Caveney's considerable skills.

I gratefully appreciate the help and encouragement of the noted UFO researcher from the United Kingdom, Timothy Good. His initial combing made a fairly good manuscript into a great one.

# INTRODUCTION

Who was Barney Barnett? He was a New Mexico soil conservation engineer and surveyor who found a crashed flying saucer with alien bodies in July, 1947. He had been with the conservation service since 1935. The government parent organization for this conservation work was the U.S. Department of Agriculture. After serving in several other similar positions in Colorado and New Mexico, he was posted to the Salado Soil and Water Conservation District in the spring of 1946. While driving south on New Mexico Highway 12 on a warm July morning, he saw metal glinting in the sunlight a short distance away. Thinking that it was a crashed plane, he found a way to it and drove over. He told a good friend that he had no qualms about it, "The craft was a vehicle from outer space." This is that story. The man, his character, his background, and the aftermath of his find and how it affected all who were associated with it.

Barney Barnett told his friend, Vern Maltais, about the incident in February of 1950. As far as we could tell, the craft came down early in the first week of July 1947. Ruth and Barney Barnett had agreed not to write about the crash to friends and relatives because of job pressures that might arise. This was where he wanted to be and had been posted in a job just a year or so earlier. He did not want to jeopardize his position or his very good reputation by claiming he had seen a flying saucer crash, which many wouldn't believe.

The author had spent some time delving into Barney's background and had located extensive materials providing insight into his early life, including college and his stint in WWI with an engineering outfit attached to the 88th Infantry Division. We were able to find a civil service application from 1940, which was immensely helpful. Barnett was valedictorian of his 1914 Humboldt, Iowa college class. Valuable information was received for this manuscript from other colleges concerning this and related incidences. We received excellent information from Harvard University, Indiana University, University of Michigan and a two-year vocational school, Oakland Polytechnic, where Barnett received his engineering training.

This book is also about physical extraterrestrial material debris scattered about the crash and the locating of this material by metal detector, and excavation. Early in this investigation occasional material was found on the surface or in the sage, or buck brush. The exterior pieces we believe we found matched the description and toughness given by Jesse Marcel, who saw the initial Roswell debris. We believe it is possible two separate crafts may have been involved from different technological timelines possibly controlled by the same alien entity. The theory that the two craft may have had a mid-air collision makes more sense if they were operating in close proximity to each other.

Every piece found we felt was important underwent a thorough scientific analysis. Sometimes two or more labs were used to get our final results. One lab found that pieces of a tiny shoe sole were made of cellulose, but only a forensic lab could tell us which kind of cellulose; there were a dozen or so common types. The shoe sole had nine layers of material and is featured in Chapter 13. We also found some gray rubber coated fabric, some wax pieces, and several items we could not identify. By far the most complete analysis was done on the metal chards. Their discovery, analysis, and possible location in the UFO are found in Chapters 4, 5, and 6. We found some 21 lbs. of this metal and analysis was begun using an ICP-MS process and other sophisticated lab equipment. This process was developed and first used in 1989 and is very expensive. The analytical devices are the size of a workbench and cost up to $1,000,000. Each analysis run cost $500. We had 11 different runs done on our material. In dealing with large scientific organizations that do lab analysis such as the ICP-MS, we had to develop a protocol and professional business relationship with five different laboratories we used. UFOs or the nature of our project were not mentioned to these scientists. Isotopic tests from these reports were compiled by a material analyst in a separate facility before being included in this book.

Later in our research after 2010 we found three large pieces that look like aluminum wasps nests and they were labeled honeycombs. We believe that this material was sandwiched between a "tough outer skin," similar to what Marcel described, and a thinner skin jut inside. Indications are that it has something to do with anti-gravity or propulsion. One of the first finds in 1995 is HDPE artifact. It was collapsed in on itself and had originally been soft. Medical experts were baffled, but believe it may be some kind of artificial body part.

We found some retired military officers who had knowledge of our crash. The late Lt. Col. Philip Corso mentions the Plains crash three times in his interviews over several years. Colonel Corso was the highest ranking military officer out of the Pentagon to come forward with information about UFO crashes, aliens and specifically our Plains crash. Another retired colonel was William Leed, who as a young lieutenant had been told about the Plains UFO crash by a superior officer in Washington D.C. in 1963. Leed first visited Barnett at Socorro in 1967 and confirmed the crash. Another retired officer was the late lieutenant colonel Jesse Marcel, Jr., who saw the UFO wreckage his father brought home late one night from the crash site to Roswell. His father, Jesse Marcel, Sr., gave an interview in 1978 where he alluded to another crash in western New Mexico, found by a surveyor. Barnett's chief work with the Soil Conservation Service was surveying in his district, which was west of Socorro, New Mexico, including the Plains of San Augustin.

One of the most incredible finds in this research was a letter from Herbert W. Dick, a 1947 Harvard Graduate student and archaeologist, to his advisor stating unequivocally that he and his party of six began a stratigraphic survey on the Plains of San Augustin on July 1st, of 1947. He further stated he was there for two weeks. What was unusual about this? Dick and others of his group were interviewed by UFO researchers denied being on the Plains at all in early July. Barney Barnett told his friend Vern Maltais and others that he met a group of archaeologists at the crash sit. The Dick party camp was just a few miles away at a site Dick was to make famous - Bat Cave. Credible witnesses, strong circumstantial evidence, and considerable physical evidence in the arroyo point to something extraordinary happening there in the summer of 1947. This book is the culmination of over a decade and a half of research into these events.

The site where the UFO was found was a study in itself. The crashed craft apparently had bounced one or more times and had split its exterior skin before coming to a rest on a small rise. It had apparently cut a wide, shallow trough and disturbed the sage and huck brush, which did not grow back; we later called this "the gap." Barnett and others reported several alien bodies when they found the craft.

It was interesting to see how the Roswell events played out and the early Roswell (and the Plains) writers' efforts to influence and publicize their denial of a two-crash scenario. It was almost amusing to see writers contributing to the Plains crash cover-up while complaining about the government doing the same thing with their Roswell efforts. One writer went on to compare Barnett's discovery to the "various stages of a modern myth." By version five, Barnett was eliminated. As if to complete the myth, two Roswell writers coauthored an article about how Barnett conjured up his "fantasy" on the Plains crash. In their efforts to give Barnett a coup de gras, they followed the modern myth scenario perfectly and tried to discredit him once and for all.

Aspects of the Gerald Anderson story appear in these pages. Gerald was a 5-yr-old in July 1947 when he and members of his family came upon the Plains crash site. Considerable communication problems occurred between Gerald and some of the early Roswell researches between 1991 and 1994. We believe Gerald was there with his family in 1947, but the communication problems with UFO researchers caused many to doubt his story. Gerald became friends with Chuck Wade (an associate) in 2008-09 and was invited on the first dig in 2011 and brought an excellent metal detector. Gerald took us to another area we had not worked, the party found large pieces of the long-buried crashed craft. He returned again in 2012 and joined a larger crew who had several additional metal detectors. More crash material was found.

The spelling of San Augustin was the original spelling of the early 4th century Catholic theologian. Old maps and geologic reports used this longer version. Modern day usage has shortened it to "Agustin." We have also used the short version when quoting the name in chapter references. Occasionally "Augustine" is used and spelled with an "e" on the end.

The presentation of material herein is as objective as we can make it. Opinions and observations expressed, unless attributed to someone else, are the author's. What is written here is a simple, yet incredible story. A story of a built-in cover-up, a story of ordinary people, and how some handled an extraordinary event. Those who witnessed the event were initially met with a huge challenge. They had to explain to others an unbelievable scenario and still maintain good relations with those people. We are grateful for all of those who have expressed confidence and have supported these research efforts. Where questions exist in our minds, we have raised them for readers to consider for themselves. Those who seek more enlightenment than these pages can provide, are directed to the numerous chapter references. It is up to you, the reader, to determine if the material presented here has relevance to an extraterrestrial UFO crash on the Plains of San Augustin.

## ABOUT THE AUTHOR

Art Campbell has been interested in the UFO phenomena since the mid-1950's. He is credited with forming a NICAP (National Investigation Committee on Aerial Phenomena) chapter in the Kansas City, Missouri area. He was actively involved in the early investigation of George Adamski, whose bogus claim of being contacted by space men in the Kansas City train yard was published in Donald Kehoe's *UFO Investigator* in 1959. A quote from Kehoe about Campbell's NICAP work appears on the back of this book.

Art began a 30-year career as an educator in 1959 and holds a BFA (Bachelor of Fine Arts) and a Master's degree in education. He has served as a teacher, counselor, coach, and high school principal in his career. Art is a veteran of the U.S. Navy, serving in naval air installations and the aircraft carrier, U.S.S. Boxer, during the Korean War. He returned to UFO work in the early 1990's and began work at the crash site featured in this book in 1994.

Art is recognized as an expert on President Eisenhower's secret trip to Holloman Air Force Base in 1955 and a possible alien contact. Mr. Campbell's investigative efforts on President Eisenhower and Air Force One was featured in Timothy Good's 2013 book *Earth, An Alien Enterprise*. He is also a noted historical writer in his home state of Oregon, having had several books published on pioneer and early settler history. Art has also written for the Oregon Historical Quarterly and his historical works were endorsed by *The Oregonian*, Oregon's leading newspaper. Art is known for his professionally done graphics and videos.

Art is available for interviews and conference presentations.

# CHAPTER 1

## THE PLAINS OF SAN AUGUSTIN

When the explorer Coronado came north from Mexico in 1540, his party was considerably west of the Plains of San Augustin. He entered the present United States boundary south of Tucson, Arizona, an area then claimed by Spain. His party eventually turned east towards the present town of Santa Fe. Some of the party traveled northeast into what is now Nebraska. Another explorer, Don Antonio de Espejo, came up the Rio Grande Valley about sixty miles east of the Plains in 1582 and laid the groundwork for the first Spanish settlement at San Gabriel in northern New Mexico.

As exploration parties moved north into New Mexico and Arizona, it is believed that side parties were sent out to the east and west. One of these parties probably first saw the Plains. Over the years Spanish conquistador equipment has been found in the Plains area. In the 1920s a sheepherder found some three-inch rowel spurs near Magdalena. At a place called Kid Springs, a solid silver bayonet was found. In 1935, an iron shoe was found, which later was determined to be part of a Spanish stirrup. Also, in the 1930s, one of the Hubbell boys found a Mexican cavalry sword near Bat Cave. It is not known exactly when the Plains was named, but it is believed by some historians that the name was bestowed in the early 1600s when the religious order of the Jesuits began missionary work in the area. The Spanish Jesuit influence was interrupted in 1767 when King Carlos of Spain deposed the Jesuit leadership. After that time the Franciscans were encouraged to continue the mission efforts. Some are still there today. The religious order of the Augustines was named after St. Augustine who lived from 365–430 A.D. The order was founded in the sixth century A.D. St. Augustine was quite influential in the philosophy and theology of the early Catholic church.

Today the Plains of San Augustin is located in west central New Mexico. U.S. Highway 60 runs across the northern one-fourth of the Plains, between Socorro and east-west to the tiny settlement of Datil, New Mexico. The Plains sits astride Socorro and Catron Counties (Map, Chapter 9, Figure 9.3).

My wife and I first visited the Plains in the early 1990s, returning from a trip to the East Coast. We had heard of this area named as the location of a possible UFO crash in the past. Since it was on our way home, we decided to drive over and have a look at the area. We contacted several people who suggested possible locations to investigate. Spending the night in Socorro, we arrived at the Plains early the following day.

At first the Plains gave the impression of a sage-encrusted barren wasteland, but we were taken by its simple beauty and wildlife. From the road we spotted several pronghorn antelope and a young eagle gliding overhead.

Morning on the Plains is a beautiful, quiet, peaceful time. The clear morning was greeted by a rosy orange glow

*Figure. 1.1. **Plains of San Augustin**, looking southeast towards Shaw Mountain. The Plains is some 59 miles long, varying in width from 11 to 19 miles. The Plains was originally a gigantic lake bed holding Ice Age melt-off. Hunters of the Folsom and Sandia complexes hunted big game here approximately 11,000 years ago. An archaeological party working here may have discovered a UFO crash in 1947.*

over the San Mateo Mountains. We parked the van and started walking. Small birds had been active for the last several hours. We flushed a covey of blue quail behind some rabbit brush. The sun was just starting to rise in the sky over the mountains to the east with enough light to cast long shadows on the colorful landscape, and put a rosy golden glow on the uppermost tips of the sagebrush. Small flocks of pipits flew and dipped near the edge of the old lake bed. We heard a horned meadowlark somewhere ahead as he guarded the mound of earth and called to others in the distance. We were not equipped to spend much time there, but in several hours time we took photographs and explored the shore of the old lake bed and some high ground.

I really do not know what lured me back to the Plains the following July, but when I found the HDPE artifact, I knew I had something unique and unusual. In the next ten years, I sent for many scientific articles concerning the Plains and area geology, archaeology, anthropology, hydrology, flora, and fauna. I read books by local people about pioneer life. I talked with town residents and merchants to get a feel for the small communities around the Plains. I scanned hundreds of old area newspapers (on microfilm) from 1935 to 1950. I talked with land owners, ranch hands, and former area residents and made contacts at a college in the area. In the succeeding years after the first visit, I returned to the Plains many more times. I logged over 25,000 miles on my van and pickup truck over the next few years. I made telephone and written contact with over 70 individuals from 14 different states and several foreign countries. Little did I know when I began this research that it would be one of the most expensive, fascinating, and time-consuming periods in my life.

The Plains has been studied scientifically since before the turn of the century. Geologists designate the Plains as a down-faulted basin later filled with a melt-water lake. The water came from the Pleistocene ice invasions of the last million years. The last ice invasion began about 70,000 years ago. It was known as the Wisconsin stage and reached its maximum coverage some 18,000 years ago. It is believed that by 9,000 B.C.E. much of the ice in the mountains had melted and the Great Basin lake was some fifty-seven miles long and fifteen miles wide.[1]

*Figure. 1.2. **The Very Large Array** (VLA) radio telescopes. There are 27 dishes, each 82 feet in diameter, located on the north end of the Plains on Highway 60 between Magdalena and Datil, NM. In 1982 a local woman who worked for a government agency was followed by a very large UFO as she drove west towards the VLA from Magdalena. She said it stayed above and behind her car, which contained her and two young children. She pulled over near here on Hwy. 60, got out and looked at the craft, which she said was well over 100 feet in diameter. When she stopped, it stopped. Then she said it accelerated over the VLA telescopes to the southeast and disappeared.*

One day while I was on the Plains with a local acquaintance, the mountains around the Basin were described to me. "Up to the north there, you can see the Bear and Gallinas Mountains, and those closer ones are the Datil Range. Over to the east here, are the San Mateos. That one there is Magdalena Mountain, practically goes into Magdalena. That continuous line of mountains to the south has several peaks. There is the O Bar O, Salvation Peak and the Pelona Mountain. That one by itself in behind there... is Coyote Peak. Behind Coyote there... the Continental Divide runs west a ways and circles the southwest end of the plains and goes north toward Wagontongue Mountain then up toward Pie Town." He turned west again into the brassy sun, put his hand up to the brim of his weather-worn hat and continued. "Thems the Mo-Gee-On Mountains (Mogollon) and due west there is the Tularosas and behind them is the San Franciscos. They go all the way to Arizona."

The Very Large Array (VLA) is one of the most impressive sights on the San Augustin Upper Plains. This is one of the world's most powerful radio telescopes and is built to synthesize a single radio telescope some twenty miles in diameter. The VLA provides radio photographs of celestial objects and signals as well as pulsars and quasars. The VLA is placed here because in this desert climate of 7,000 feet, there is low water vapor and a minimum of blurring of the radio signals by clouds. The surrounding mountains block out earthly interference from radio and television transmissions. The large, flat upper Plains of San Augustin helped with the easy movement of the 27 radio telescopes on over 13 miles of railroad track on each of the three arms.

In 1972 Congress authorized construction and appropriated the initial funding. The first of the 27 telescopes, weighing 230 tons, was finished in 1975; the last one in 1979. Scientists all over the world visit and work at the VLA. Today it is administered by the National Radio Astronomy Observatory (NRAO). The cost at completion of this large project was $78.6 million. A very nice visitors center is open most days of the week. Some of the movie *Contact*, starring actress Jodie Foster, was filmed there in the early 1990s.

In the mid-1950s, about eight miles southeast of the little hamlet of Horse Springs, a 2,000-foot cored hole was drilled. This is known as the Oberlin Hole. From this and other core holes drilled, scientists are able to get a better picture of the geologic and hydraulic features of the Plains of San Augustin.

A study of the pollen from several deep holes suggests a cool moist temperate forest climate some 18,000 years ago and before the first melt lake was formed. The eventual change was from a cool and moist climate to the semi-arid climate we have today. Although the lake disappeared three to four thousand years ago, the rich marsh and grassy receding shoreline attracted large game. Fossil remains indicate past climatic conditions and some of the Plains' environmental aspects. Surface finds in the basin include extinct horse, bison, bear, and what may be fossils of the mastodon, as well as many other smaller mammals, birds, and fish.[2]

Langford Johnston in his book, *Old Magdalena Cow Town*, relates that a bunch of cowboys were exploring a fissure on the Plains about 1930. One went down into the fissure and emerged with a "mastodon tooth" about ten inches high and eight or nine inches across.[3] It was said to have been used as a doorstop at a saloon in Magdalena for years.

Several writers have noted that much of the extinct lake has everything but water. It has shorelines, cliffs cut by wave action, beaches, spits, bays, bay mouth bars, and islands, one of which fits prominently into this story. It was on the shores of the lake bed to the northeast that evidence of early man has been found dating back about 11,000 years. Some of these shorelines are also known as terraces, which will also be discussed later.

Two other deep holes for oil exploration were drilled in the 1950s and 1960s.[4] One test hole reached 2,000 feet and another penetrated to the 1,200-foot level. Thirty-one known water wells had been put down into the lake sediment for stock watering purposes by 1974. Water was found as shallow as 50 feet and as deep as 500 feet. Many of the original wooden windmills have been replaced with steel. Here and there water is now pumped by solar power.

From additional research, we have learned there are several vegetative zones in and around the Plains area, and that the vegetation varies according to the elevations. In the valleys we find desert grasses, mesquite, some varieties of cacti. Near arroyos and streams, salt brush and creosote bush grow. Many of the desert grasses such as gama, three-awn, tobosco and salt grass, provide livestock forage on the Plains. This is generally called the lower Sonoran Zone. Above the grazing lands can be found forests of piñon, juniper, pine and scrub oak. Higher can be found aspen species, ponderosa pine, and Douglas fir. In the entire Plains basin and surrounding hills in the spring and summer, can be found many varieties of wild flowers.

*Figure 1.3.* ***Picturesque windmill*** *such as this one with old loading pens can be seen along highways around the Plains. Wooden windmills are still in the area, but mostly replaced by metal towers. Electricity now pumps most of the water. Solar panels mounted on the towers are furnishing some of the power. Photo taken on the Rob Emory ranch south of Datil.*

**THE BIRTH OF AN ARROYO**
High in the windswept mountains east of the Continental Divide on an old eroded bench, stands the remnants of a grove of large Douglas fir trees. All that is left now is a series of deep gullies and some stumps with protruding roots and rocks, which have separated from exposed rock ledges. The trees were logged sometime before 1900 by hand axe and crosscut saw. This soil has been disappearing now for a century or more. What can be seen now is a wasteland of highly eroded ground, marked here and there by the weathered, whitened stumps, highly reminiscent of a hillside of gravestones.

The once-magnificent trees supported a canopy of branches that protected the ground cover on this part of the ridge. As part of the cycle, a light, but cold rain falls, hardly soaking into the already saturated ground. Some of the water on this side of the mountain drainage divide started its journey here. Some of it was taken off by a wind before it fell, and it probably became a part of the evaporation process. Most of it, however, fell to the wet soggy ground and began its run-off cycle.

Rocks have fractured off parent ledges from the freezing and thawing process, and the tired old roots seem no longer able to hold them back. The weathering and breakup has been aided in this disintegration process by some hardy plants on the slope. Cracks have appeared in the rocks. Roots and fibers were thrust into crevices, and the rocks broke up even more. Acids were formed and assisted in the decomposition. A thousand summers and winters have done their work on this mountain, grain by grain, pebble by pebble. The down-slope movement of the mountain top materials has worked its way towards the lower elevations.

The terrain below the old weathered stumps has been well defined. All of the ground now slopes into gullies and washes, and the rain wash flows in them on a predetermined journey. At first the slope is gentle, and the water pools behind the rocks, then flows around and then over the rocks into the ever-deepening gullies. Below the old stumps several gullies merge into a stream, and the brown water, now deepening, picks up speed as the slope increases. Now the stream runs over bedrock and large rocks that have been undercut in the flow path. Deflected by these resistant obstacles, it swings away, undaunted, and continues the downward journey. Below is a confluence with another stream coming from the left, and the two become one.

The brown churning water is now almost a foot deep. The gradient becomes more gentle. The water starts to pool and cascade at intervals as it reaches a brushy area above an aspen grove. It runs under a tangle of old roots and a log, then disappears into the thicket. Several springs made up of subsurface ground water from the saturated soil above join the creek as it winds and twists through a deep channel in the grove, now slowed by logs, leaves, roots, and debris from the trees and undergrowth in its path.

Below the aspen grove is a meadow, which has a much gentler terrain, then, as it flattens out, the water flows into a long pool and continues on slowly over a breached beaver dam. Below the meadow is a stand of piñon trees. Scrub oak roots and rocks have now slowed the water considerably. On top of the banks of the stream it is lush and green. Grasses grow and wild flowers will bloom when the weather warms up.

Now there is less volume, as some of the water has seeped into the creek bed, which is floored with rock waste and silt deposited by the stream in years past. Below is a steep bank, flanked by juniper trees, that leads to a sandy arroyo. A well-cut notch lets the water onto the sand. The water pools for a short distance and does not go any further... it has reached its base level, the lowest point this water course can erode.

The water here has pooled because of a check dam across the arroyo about four hundred yards downstream. Several of these check dams, above and below, greatly slow the water while protecting the valuable grazing land on both sides of the arroyo from erosion and excessive run-off. Like the beaver dam above, the dam helps stabilize the banks and provide a water source for livestock in the dry months of May, June, and July.

The arroyo has hundreds of small water courses emptying into it. In some years there may be little rainfall on this side of the mountain. But a heavy snow melt higher up or thunderstorms somewhere else generally means life-giving water behind this check dam and others.

We are still some eight miles from the Plains, and our arroyo course will meander through the valley, picking up pyroclastic material known as volcanic ash, lava, many types of rocks, and other organic substances. During their transit down the mountain side, the pebbles left from the original rocks were less coherent and softer and were

*Figure 1.4. **Plains from Bursum Rd.** Looking northwest towards Crosby Mountain. For over 125 years the Plains have been prime grazing grounds for cattle. Many ranchers own and lease land from State and Federal sources as well as each other. The Plains are interspersed with sandy areas called playas. A large playa can be seen in the distance.*

the first to be reduced by abrasion to sand. Silt and mud came down also with the sand and are the finest-grained products of erosion in the arroyo. The lighter material stays on top in the arroyo which may be 15 to 30 feet deep. Below the silt and sand are gravels, pebbles, and cobbles, in that order. At the base are boulders (10 inches or more in diameter). All of these abrasive substances move in the arroyo at different rates, constantly rubbing, knocking, scraping, and grinding each other. In some years these abrasive neighbors hardly move at all. In times of high water and flood, these destructive elements work together, fed by countless rivulets and thousands of streams, each bringing down, rock by rock, grain by grain, the transplanted mountainside into the arroyo and to its alluvial fan below.

# CHAPTER 1
## REFERENCES

1. Webber, Robert H.; Report; "Pluvial Lakes of the San Augustin"; Guidebook; 45th Annual Field Conference, Mogollon Slope; West-Central New Mexico; New Mexico Geological Society, 1994.

2. Powers, William E.; Abstract; "Basin and Shore Features of the Extinct Lake San Augustin, NM; *Journal of Geomorphology*; Vol. 2; New York, NY; Columbia University; 1939; p. 207-209.

3. Johnston, Langford H.; *Old Magdalena, Cowtown*; Bandor Log, Inc.; Magdalena, NM.

4. Corie, Edwin; *The Gila River of the Southwest*; The University of Nebraska Press; 1964; p. 87.

*Several excellent works concerning the geology, mineral resouces, and hydrology of the plains were consulted during this search.*

> H'addril, Marilyn; "Listeners of the Universe, New Mexico's Very Large Array of Radio Antennas"; *Friendly Exchange Magazine*; Spring 1993; p. 30-31
>
> "The VLA Pamphlet"; The National Radio Astronomy Observatory Facility of the National Sciences Foundation; Associated Universities.
>
> Powers, William E.; Paper; "The Volcanic Rocks of the Western San Augustin Plains District, New Mexico"; 1939.
>
> Cibola National Forest Map; Magdalena Ranger District; US Dept. of Agriculture; 1991.

# CHAPTER 2

## THE SITE

With the landowner's cooperation, we marked the gap area with one-foot wide aerial flagging tape and took pictures from different angles as we flew over the arroyo. The basin was about 300 yards wide at the lower end where we were working. Over thousands of years the material that washed in from the mountains, called alluvium, spread out as the valley widened. This is known as an alluvial fan. These alluvial fans are made of silt, sand, gravel, pebbles, cobbles, and boulders, in that order, as mentioned in the previous chapter. Considering the contours of the high ground on the edges of the alluvial fan, we estimate the depth of the fan to be 35 feet to 50 feet deep. As long as we were flying over, we decided to fly about five miles directly in line from both ends of the gap. We thought that if something had come in here, it might have left skip marks or other ground disturbances. We found nothing.

For the purposes of clarification, we will call the wide alluvial fan the arroyo. The old main water channel, the old water course, and those channels that run more or less parallel through the arroyo, we will call water courses. The area we will be discussing in some detail will be referred to as "the gap" or "bare area."

In past years when there was much more water going into the arroyo basin and the flow was not checked, the alluvium was loaded with fine sediment. As the arroyo basin widened, the old water courses were supplied with more debris and silt than they could carry away. As the arroyo neared the Plains, this silt and debris continually clogged the old water course channels. What water there was from above lost much of its volume through seepage and evaporation. Check dams were put in for conservation purposes over the years. In past times of higher water flow, these clogged channels in the alluvial fan became blocked and new channels were made. Occasionally a channel that contained a runoff water flow would become blocked, and the water, seeking the path of least resistance, would build up, eventually crossing to another water course. Geologists call this a braided streambed. For the first few years or so of the project, we assumed that the bare area or gap, was an old water channel, cross-over braid between two old water courses.

About fifty yards above the arroyo braids, the old main water course is clogged by silt, sand, and old debris dating back to the 1930s. This was quite a mystery, because all of the old material dating back that far should not have been on top of the ground, but below it a foot or more. This was our first indication that the ground had been disturbed. When the head of the braid was clogged by silt, this shut off all the other water courses in the braid. The water within recent years started a new channel above the blockage, and all arroyo run-off is channeled now in one braid where it runs a hundred yards or so before soaking into the porous soil.

The bare area or gap looked much different through binoculars from an old island some distance away. It looked very much like a section of an old wagon road. It would not have attracted much attention, except for one thing. The edges were defined on both sides by buck brush. This gap was about 135 feet long and some 28 feet wide. The buck brush is also known as black brush, with a scientific name of *coleogyne ramosissima*. The gap, whatever had caused it, slashed across the end of a strand of this brush between two arroyo braids. Ninety-eight percent of what should have been brush in the gap appeared to be gone! When I took soil samples, 21 test holes were dug to a depth of about 12 inches (30 cm). Some light gravel was found, but no larger material.

There were other items of significance. Other channels, which had held meandering water flows, were irregular. The bare area was very wide with no sign of having had water running through it. For the water to have made this bare area, the main old water course would have to have filled with silt to over one-and-a-half feet in depth, channeling water into the area. There was no indication that this had ever happened. If it had, there would have been some signs of a channel there. All of the braids that lead to the Plains had signs of channels in them. The diagonal bare area did not. The final bit of information was perhaps the strangest — the water would have had to run uphill to go through or over the bare area, which is on elevated ground. Once we realized that the bare area was where all of the small material from the crash was located, it became more obvious to us that this was a swath through the buck brush made by the crashing craft.

The accumulation of cans and bottles a mile or so below highway culverts and bridges was fairly common until recently. It was easier for those of lesser means in the area to throw accumulated refuse in the arroyo and let the next flash flood dispose of it. Out of sight; out of mind. Today's more affluent ranchers usually have dumps on less conspicuous parts of their property where they can take refuse and use farm equipment like front loaders and backhoes to bury it if necessary. Dead and diseased cattle are buried in this way.

When a flash flood comes into the Plains area, the farther the water gets away from highways and small bridges, the less channel it makes. During a flash flood, an active arroyo can be four or five feet deep where a road or bridge crosses over it. A mile downstream, the water slows down and spreads out into a delta near the channel braids. As mentioned before, old trash and wood debris washes to the head of a braid, completely blocking the braid and water is forced to find another way around it as it continues toward the Plains.

Most of the early interesting surface material I found was after a rain. The hammering force of the short, rather violent rain storms exposed things in the gap and in the old water courses. On one occasion after such a rain, I found three lumps of foil sticking out of the sandy soil that were not visible several hours before. The process seems to reverse itself in the hot, dry, windy weather in April and May, when blowing dust and silt covers over what may have been exposed before. The water seems to be winning out. More has been exposed than covered up. The visibility of the ground in the gap earlier in the spring is not as good as it is in July and August, when the edible forage is grazed off.

**RATTLESNAKES**

A cautionary word needs to be said about rattlesnakes. The area is not infested with them, but they occasionally make their presence known. Many old prairie dog complexes in the arroyo are used by snakes for two reasons. They need to be out of the sun and out of sight as protection against predators, mainly hawks and eagles. Some kangaroo rats use these old prairie dog complexes for the same reasons. This also furnishes food for the snakes. Both are nocturnal, staying in the holes during the day and foraging at night.

One afternoon after a hard rainstorm, I walked in the old water course looking for foil. I noticed a few kangaroo rats darting around and did not think much of it, until I noticed several rattlesnakes also above ground. Then I realized that the prairie dog holes had been flooded. This had forced both out on the surface. I do not fear rattlesnakes, but I have a great respect for them.

I had one other encounter while at the site. One day while taking a break, I noticed a large rattlesnake working its way towards my excavation. I tried to shoo it away with shovels full of dirt and rocks. It would stop briefly, but kept moving towards the hole. I had to kill it, as my digging had cut through some of the old intact tunnels of the prairie dog complex. I had been working in the excavation on my hands and knees, and if the snake had made it into one of those tunnels, I could not continue my work. Later I skinned the snake and hung it out on a juniper snag to dry. It can be seen in the background of the photo in Figure 2.3.

In a letter to the author, Vern Maltais brought up another interesting snake episode. Vern and his wife Jean were longtime friends of Barney Barnett, who was the first witness associated with the Plains crash. It seems that during World War II, he and Barney Barnett were out in Barney's pick-up truck near Mosquero, New Mexico, some miles northeast of the Plains. Maltais said they "saw a large sidewinder rattlesnake and stopped. Barney took his surveyor's spade and had the snake strike against the spade several times. After several strikes he cut the snake's head off and opened the snake's mouth to show me the fangs." (Barney apparently put a small stick in the snake's mouth to keep it open.) The stick slipped and glands from the snake's head spewed venom all over Barney's face and glasses. Maltais said, "I just ducked out of the way. We were out in sort of desert country and fifteen miles from water. We were both white."

Living in snake country, one learns to take certain precautions. Among the things I do is be ever-vigilant and know the dangers, wear good boots, and never put hands, feet, or legs where I cannot see first.

## HANTAVIRUS

Throughout the western United States, rodents carry a deadly disease known as the hantavirus. This virus comes from fleas found on deer mice and other rodents. The author wishes to warn the researcher to be extremely careful around packrat nests, often visible above ground, on or near fallen trees, and root outcroppings.

These nests of little chips of wood are constructed by several varieties of kangaroo rats. Many people have died as a result of exposure to the fleas on these rodents, especially in enclosed areas under houses or sheds. Breathing the dust from the feces can cause severe respiratory and other fatal symptoms

*Figure. 2.1. As first seen in 1994. A detailed analysis of why the buck brush was absent (right) was performed using the help of a soil conservationist and soil analysis from the NM School of Mine Technology. This was later called "the gap."*

*Figure. 2.2. **Chuck Wade with his nephew Jason on 2005 dig** setting up and calibrating survey equipment. Folded foil shards were found buried here in 1999. The laser leveling equipment points to a line of cliffs 10 miles away where the archaeologists camped at Bat Cave in early July 1947.*

similar to the bubonic plague of the Middle Ages. There is no known antidote for this disease, which often ends in death.

## THE SOIL ANALYSIS

The post-hole digger I had rented that day made a kind of crunching thud sound as I started digging the soil samples. I had divided the long 135-foot long gap area through the buck brush into three lengthwise sections. Seven of the test holes were in the center of the bare area of the gap; seven went through the brush on the right, with an additional seven through the brush on the left. What I wanted were soil samples from where the brush was growing on both sides and a sampling of soil from the center where hardly any was growing. This was a sort of experiment with two rows of control samples and one experimental row of samples for comparison. While in Socorro, I had rented the post-hole digger and bought a case of half-pint canning jars. I dug the test holes the depth of the post-hole digger and took the 21 samples from that depth.

If there was anything in that arroyo basin soil to keep the buck brush from growing, I wanted to know about it. I finished just after it got dark and managed to get the holes filled in, as I had promised the rancher, so the livestock would not step in them. I had to retrieve the last several jars in the brush by flashlight. Earlier, I had gone over every foot of the gap area, including an additional 50 feet on each end with a Geiger counter. I had borrowed the U.S. Government device, which was a scan with a Victoreen 290 survey meter using a 489-55 scintillation probe. As I was getting instructions on using it, I was told that any audio sound and meter reading that was over four or so would be significant.

That day I had spent in Socorro and the sun was setting when I got back to the site. I remember thinking, here I am out in the middle of nowhere, probably standing in a deadly radioactive area and this may be my last sunset. I need not have worried. The survey meter hummed right along and did not register over one or two, which was a normal background reading. I had to get the Geiger counter back the next day. While walking back up a small rise, I found some interesting little rotted textured, dried rubber pieces, which I picked up. I also found what appeared to be an old child's shoe sole, on lower ground.

The agricultural testing laboratory had previously told me how to gather the soil samples. Essentially, they asked me to send three samples. I was to take one-half cup of soil from each of the seven jars of soil from outside the right-hand side of the gap and where brush was growing, and mix them. I did the same for the jars from outside the gap in the brush on the left side of the gap and mixed them. The same was done for the seven jars from the center of the gap. When I was ready to mail them off, I had three plastic bags of soil, each mixed only with soil from that row. The laboratory labeled the field samples:

- #9732 (from right side of gap),
- #9733 (from the center of the gap), and
- #9734 (from the left side of the gap).

What we discovered can be seen below.

The next time I went back to the site, I spent some time looking at the mountains that surrounded the Plains. I

*Figure 2.3. **Author** taking notes in daily journal in 1997. The rattlesnake skin in background was made from the snake he killed a day before. Snakes were not a problem at the site, but one exercised caution.*

looked back at the high windswept mountain to the north, stooped down and picked up a handful of the damp, sandy earth. It had rained the night before and it smelled good. I began to let it trickle slowly between my fingers. So here I was, where the mountain top came to the Plains! Here in my hand were the three major soil-building blocks: phosphorus, potassium, and nitrogen. There were other elements trickling through my hand, magnesium in balance with the calcium. There was sulfur, boron, and zinc. A cool breeze had started to come up and a rain cloud overhead looked threatening. Still, trickling out slowly, the heavier sandy soil dropped down and the lighter loam was picked up by the wind and was blowing away like a wheat chaff. With the lighter material fluttering in the breeze was a tiny piece of foil, probably left from someone's gum wrapper (we surmised). The soil I held still contained some sodium, iron, and copper. Then it was gone. Still left in my hand was some organic matter, a tiny root and a rotting piece of wood. Could these have been from the old Douglas fir stumps or perhaps from the aspen grove? I let them go and they returned to the soil at my feet. Wow! I thought as I rubbed my hands together to get the dust off. Dust to dust, ashes to ashes and life goes ever on. Somewhere overhead I heard a crow call and at that moment I felt a drop of rain. The process was continuing. Little did I know the importance of that piece of foil. It was not from a gum wrapper. We now believe it had been made on another planet.

When the report from the soil lab came back, we found that the soil on the right side of the gap and the soil on the left side of the gap were notably different than the soil from the center. Soil is far more complex than one might expect. When you

*Figure 2.4* **The 2004 Navajo Crew** *worked tirelessly on this dig. Twenty-eight test holes a meter and a half square and 20–30 centimeters deep were made. Much of Chuck Wade's original foil was uncovered (analysis in Chapter 4) in the four days at the site. Left to right: Benjamin Shirley, Merle Lawrence, Dick Shirley, and crew chief and aspiring journalist Mervyn (Murff) Tilden. Benjamin found the first piece of the larger metal shards, which is believed to be from the exterior of the craft (see the photos in Chapter 4.) Murff was with us for 11 years or more, on digs in 2004, 2005, 2011, and 2012. In the meantime he graduated from college and was doing graduate work as this book was being published.*

*Figure 2.5.* **Ice Age Pleistocene lake basin** *from 11,000 years ago. Lake at one time some 5,000 years or so ago held glacier run-off. No major bodies of water came into the lake and it eventually evaporated. Ancient Chiricahua Indians lived along the shores. Evidence of first maize (corn) found in caves goes back some 3,200 years. In the early years, weather patterns caused a lot of Native American artifacts to become exposed. This also contributed to the crash artifacts being found near the surface. When metal detectors were brought to the site, many metallic artifacts were found. There was evidence of the site being monitored for some fifteen to twenty years after the crash was first found.*

*Figure 2.6. **Bobby Wade** accompanied Chuck and the Navajo crew to the Plains in 2004. He was an enthusiastic worker who helped set up the commissary area. His help was much appreciated.*

consider that all the elements in the arroyo are concentrations of the surrounding mountains, hills, and rangeland, the number of elements found in that arroyo were astounding. Some of the elements found include concentrations of phosphorous, potassium, manganese, nitrogen, sulfur, boron, and zinc. The report determined the pH reading (measuring acidity) was the same across all areas of the gap. Soluble salt content was also recorded and determined to be significantly lower in the center section. In addition, the report measured organic material, which is generally recognizable as something that had been growing, but has returned to the soil. The soil report indicated that the right side of the gap and the left side were identical. The center of the gap had the same materials as the left and right, but the percentages were skewed for the organic material and some of the other elements.[1] The question arose: why was the center of the gap, not 30 feet away, different from the samples we took on the edges?

I consulted a friend who was working in the soil science field in eastern Oregon. He said they often see the soil change when samples are taken from virgin ground (along a fence line) compared to samples in the same field taken a few feet away, where crops were growing. He said the soil itself does not change, but the addition of agro-chemicals, weed control, and continued plowing or tilling can affect soils a few yards apart. He looked at the soil report and noted that no unusual additions or contaminants were detected. He thought the soil in the entire 135-foot length of the gap had been turned over or churned with some type of equipment. He concluded, "Soil in the center of the gap had been added or brought in by means other than natural."

He asked if there was an accumulation of the same type of soil or sand nearby. As I started to say, "no," I remembered the high ground that blocked the bottom of the arroyo's main channel, causing the braids to branch off. This blockage I later measured to be about 30 feet long and 15 feet wide. This blockage was the reason that various braids over the years had defined their way around to lower ground. The friend said, "It looks and tests very much like fill from nearby." That was it! We had learned from some reports that the UFO crash had left a wide, shallow trench. The Army had apparently taken fill from the build-up area fifty or sixty yards away to level out the low ground in the trench. This also explained another mystery. Some human metal trash was in the arroyo, but some of the older trash was on top of the ground, when it should have been buried at least a foot or more in the sandy loam soil. We found some very old beer cans with a cone-shaped top, circa 1935. We found some steel beer cans with aluminum tops (easy to open with an old triangular-shaped church key opener.)

We found fruit cans with no edge suitable for a can opener (obviously crudely opened with a hunting knife) circa 1900 to 1922. There were some 8-ounce condensed milk cans with two small slits for pouring. There was no powerline electricity in the area until 1955–'56. The cans of condensed milk were advertised and used up to the mid-50s or so for baby formula and weaning. Adults also used them for milk in coffee or tea. Unopened condensed milk cans of those days had a shelf life of about 15 months. Most were the Borden and Eagle brands. All of these cans including a DDT bug bomb canister (banned in the 1970s) should not be on top of the ground. The fact that they were indicated that soil had been brought in from elsewhere, and was turned over with some piece of equipment, mixing in the can artifacts. We believe all were at one time covered with loose arroyo sand or soil, but had become exposed, like some of the artifacts, to rain and wind erosion when we found them.

## CHUCK WADE

We were invited to do some speaking just after the turn of the millennium, and we met Chuck Wade at the 2004 Aztec UFO Conference. The conference was friendly and small enough to be personal. They had divided the speakers up on Friday evening and assigned each to a table. We were to chat with conference people and answer questions. Chuck spent some time at the table and talking to me in private. Here was a man with a great deal of energy and as open and enthusiastic as they came. I took an immediate liking to him. He was someone who had a great deal of interest in the project and the material that I had found.

Chuck graduated from New Mexico State University in 1968. Chuck and his wife Nancy have been in Gallup, New Mexico, for about thirty years and raised a family there. Chuck also has a Roswell/Corona connection. His father owned Wade's Bar in Corona. His dad knew Mack Brazel, who found the first wreckage of the Roswell crash in early July 1947. Chuck also knew his sons Bill and Vernon Brazel, as well as others connected with the Roswell/Corona UFO crash. Although barely in elementary school in 1947–48, Chuck grew up with the basic knowledge that big things happened in those years. Chuck was eventually to bring new dimensions to the research by bringing fresh approaches to the site.

## CHAPTER 2
### REFERENCES

1    Campbell, Art. Correspondence and site soil report from Agri-check, Inc., Umatilla, OR, 11/15/1997,

*The references below primarily deal with the deadly hantavirus*

    U.S. National Library of Medicine and National Institute of Health; *Hantavirus*; Web 28 May, 2010.

    National Library of Medicine; *Hantavirus*; Web 15 June, 2010.

    "Plague in Arizona"; Pamphlet; Publication of Arizona Department of Health Services; Vector-Borne and Zoonotic Diseases Section; Phoenix, Arizona.

# CHAPTER 3

## A CROSS MARKED THE SPOT & THE FIRST DIG

I had passed it a dozen times in two years and always thought...just another two sticks wired together from an old fence or something. One day, passing by again after a rain, I caught a glimpse of something red and partially buried. I got out my whisk broom and started brushing away the sand. The object was large and looked like plastic. While clearing away the sand and dirt, I had to move the sticks. One was deliberately notched (Figure 3.2). I thought, "My God...this is a crude cross." I finished excavating the red plastic that seemed to be a child's rain jacket and put it between some folded fine wire screen I had brought along and slipped the whole thing under the steel cot I had in the truck. I was due to interview several people who had lived in the area that evening, so I took some photographs and decided to wait a day or so before continuing any more excavation.

As I drove to Reserve, about 70 miles southeast of Datil, I crossed over some 20 or more arroyos and culverts. I had been reading in the old newspapers about flash floods washing out the old roads. I could see how a flash flood over the Plains roads could have threatened vehicles, animals, and children if they got too close to these treacherous waters. Was a pet buried under the cross? Had I stumbled upon some kind of grave? Were the cross and jacket connected? Could this be a little memorial, and if so, to whom or what?

After a very good chef's salad at a restaurant in Reserve, I called the residence of the two sisters whose names I had been given. They had lived on the edge of the Plains in the late 1940s. No one answered the phone, so I drove out to their place. As I parked the truck, I saw three people in the backyard working on a garden. At first I thought I saw a woman and two children, but as I greeted them and went closer, the two smaller ones turned out to be very small Hispanic women about 70 years of age, who were the sisters, and the other adult, a married daughter. I believe the sisters understood English, but chose to communicate in Spanish through the daughter. Just after World War II they had been in their twenties living with their parents at the edge of the Plains. I asked them about a plane crash up near the town of Horse Springs. Both remembered it. (Chapter 15). I broached the question about any children drowning or a little memorial to a child. Neither knew anything about it and they did not recall any serious child mishaps. I remember thinking, "that's a relief." My real fear was that I would be disturbing an infant grave. Many of the area people are Hispanic and a still born child, if conceived out of wedlock, in those days could have been buried in such a remote place.

As I began to talk with the women, I was impressed by their healthy appearance and lucid answers to my questions. One seemed to act as a spokesperson, while the other leaned on a shovel handle not much taller than she was. When a question was asked, the one that answered would always confer politely with her sister. Nods, a few words, or just a twinkle would suffice before I received an answer. They had grown up in a rented house with no electricity, plumbing or running water. I later discovered that this was not unusual for the many Hispanic families in the area. Also, due to the isolation of the area and the lack of good roads, many had not ridden in an automobile until the late 1930s. The ladies were polite and seemed at ease with my questions. I understood that they had been interviewed by at least one UFO writer, but I did not bring the subject up. I have learned that western people are far more cooperative if they are not asked blunt questions. If I needed to know something, I could get better answers at a later time, once I had built some trust and a good relationship.

*Figure. 3.1. Cross close-up. One of the greatest surprises at the site was finding the cross. It is constructed of two pieces of juniper wood, well weathered. It is put together with 12-1/2 gauge baling wire which at one time which had been twisted tight. There were other indications that visitors had been at the site in the distant past. Note rodent hole behind long part of cross. Hoarded items below the cross may be from rodent activity (pack rats).*

I finally felt free to go ahead and excavate the site. When I returned to the arroyo the next day, I was ready to dig. I had borrowed a quarter-inch hardware-cloth screen from a rancher and took it to the site. I staked out a one-meter square in each direction. One side of the square went into the edge of an old prairie dog complex (Figure 4.8). I began the excavation. Just under the surface I found the pieces of a large bottle and glass shards with rainbow hues and several very small bottles with wide necks (Figure 3.3). The sandy dirt was loose and easy digging. I used a whisk broom to sweep the dirt onto the screen. I proceeded this way until I had a square hole a meter on each side and about 30.5 centimeters or a foot deep. After several hours I had a large pile of fine dirt under the screen and a pile of small rocks and pebbles I had dumped off periodically. When the dirt become more compacted, I scraped it off, one thin layer at a time, and swept it into the dust pan. Whenever anything turned up in the pit, as I grew to call it, I brushed away the dirt and sand around it with a whisk broom. I took rough measurements of the location and depth of the object. Nothing was as exciting to find as the bone. My journal read: "Found a yellow bone 9 ½ cm long and 2 ½ cm across. Has joint flair at top and is 6 cm in circumference. It had been cut off at small end, not splintered."

More objects were appearing in the screen. We found bottle caps and brown glass, believed to be from beer bottles. Then the T-bar appeared on the screen. I had been getting some charcoal throughout the pit. Two more bones appeared and the remains of a flashlight battery. A few inches from the battery, a pointed carbon-magneto stick came up, then three bones. My excitement over the bones overshadowed the other items. As it turned out, the foil, which I was to find much more of, was more significant than most of the other material. ICP-MS analysis of some isotopes in the foil indicated it was unlikely that it came from earth. A forensic bone specialist later identified the large bone as a part of a deer femur (probably from a poached deer). The other two bones were vertebrae, identified as being from the canine family, probably coyotes. What was significant about the charcoal and the bottle caps, was that a fire had been made under the cross and beverages consumed there. These items could not have been brought in by water, because water tends to disperse things not concentrate them.

Three kinds of lichen on the cross were identified later by a BLM (Bureau of Land Management) botanist as *xantoparmelia, ecahora muralis,* and *caloplaca holocarpa.*[1] All three are common in the west-central New Mexico desert habitats. The *caloplaca holocarpa*, one expert said, usually grows on rocks, but it occasionally chooses a hardwood surface. Juniper, a hardwood,

*Figure 3.3. (left)* **Milk bottle bottom** *numbers and letters, helped trace the time frame bottle may have arrived in arroyo (assuming it had been discarded.) It may have first been placed under the cross as a vase for wild flowers. Bottle was made in Saugus, California in 1955. Also, the rusty bottle caps were together under the cross, uphill from any water course, indicating bottles were opened near the excavation. Charcoal found in the pit indicated one or more fires at the site.*

was used in the construction of the cross. Several of these lichens can be seen on the cross (Figures 3.1 and 3.2). Since the cross had no lichen on the back side next to the ground, chances are the lichen growth started while the cross was lying face up on the surface.

The *caloplaca holocarpa*, which is rust colored, grows very slowly in a desert environment, depending upon the available moisture. Weather records kept near the Plains up until 1941 indicate that over a 26-year period the average rainfall was 13.32 inches per year. The average mid-summer temperature was 66.8 degrees and the average mid-winter temperature was 27.6 degrees. There are over 17,000 known lichens in the world, and a lichen dating of plus or minus 10 years helps historians and geologists date disturbed ground and geological movements. Without scientific instruments and certain chemical analysis, it is very hard to date this lichen. I did combine the estimates of several experts and came up with a general age of the lichen at about twenty-five to forty years. I believe the cross had lain on the ground face up for at least the entire time of the lichen growth.

There were other intriguing aspects to the cross and what was buried beneath it. The two cross pieces of wood were wired together with twelve-and-one-half gauge baling wire. The wire was twisted together with a pair of dykes (a term for a western fence tool), by someone who was experienced in working with wire. The cross-piece at one time was wired tightly to the longer piece. Over the years as the wood weathered, its surface area shrank, leaving a three-fourths of an inch gap or so between the two pieces. Several I spoke with thought it would take twenty to twenty-five years for hard juniper wood to weather like this.

When I first started digging the pit, I thought that the juniper base of the cross had rotted off and I would find another piece of wood in the ground; that was not the case. Upon closer inspection of the base, I could see that it had been squared off and the weathering pattern was consistent with the top part of the cross.

*Figure. 3.2. **Carved notch** on cross bar shows some 20–25 years of weathering. Lichen on cross bar is* xantoparmelia, *which grows about 3–4 mm per year. This helped date cross. Cross had lain, lichen side up, for 25–35 years. Absence of lichen on bottom side of cross indicated it had been undisturbed many years.*

If it had been thrust in the ground, it was not deep and probably not meant to be a permanent marker. The base also had a growth of *xantoparmelia* lichen on the end, which would not have grown underground.

**STRANGE MATERIALS FOUND**

One other aspect of the cross was a charred end of the crosspiece; it had clearly lain in a fire. As I went deeper into the pit, I kept coming up with charcoal, broken glass and bottle caps. The charcoal seemed to be concentrated near the left center of the pit and went to a depth of some twenty centimeters. An unburned wood chip also came up on the screen. There was a diagonal cut on each end, which I associated with chopped wood for the fire, or a piece from the cross construction. When Chuck Wade and Nancy started working with me at the site in 2004, they took the cross with them back to Gallup, some 100 miles north of the Plains. The lichen on the cross had a better chance of survival there in a similar climate.

The remains of the large bottle bottom found just under the surface near the jacket seemed to have a life of its own. As a sometimes-collector of Western memorabilia and some bottles, I realized that the numbered codes on the bottom of the bottle might help me trace its origin. After the Pure Food and Drug Acts of 1906 and 1912, all companies that sold bottled products for human consumption were required by law to identify their bottles.[2] I was able, therefore, to trace the brand of the bottle to the Thatcher Manufacturing Company of Elmira, New York.[3] What I was apparently dealing with was a milk bottle. Literature sent to me indicated that a plant making these bottles was opened in Saugus, California, in 1954. Other numbers on the bottom indicated the mold and job numbers. But the real surprise was a number indicating the bottle had been made at the Saugus Plant in 1955 (Figure 3.3).

While going through the *Socorro Chieftain*s of 1948, I ran onto an item commending a Datil merchant for his fresh bottled milk. Rural electrification worked its way, after World War II, from Socorro to Magdalena and eventually to Datil, Pie Town to the west, and Horse Springs and Reserve to the southeast. Electricity of this type would have been required to keep milk fresh. The Navopache Electric Co-op had drawn up their rural electrification plans during World War II, and late in 1947 had begun setting poles and doing wiring. It is believed that during those years fresh milk delivery did not reach much beyond Datil, which was 62 miles west of Socorro. The 1947 Socorro, New Mexico, phone book listed the nearest dairy as the Socorro Dairy on East Manzanares Street. This dairy also delivered to grocery stores. It is likely that the store at Datil received deliveries from them as well.

According to the manufacturer's literature, the average milk bottle made some forty-one trips during its recycling life. It is possible that this bottle had a life of some five years. A five cent deposit in those years would have kept the bottle circulating until

*Figure 3.4. **First foil (center top)**. Over the next several years much more of this was to show up. Much of what I found was just under the sandy surface and would show up after rain. Items from pit starting at bottom and going left are buttons and scrap from child's red rain jacket (earthly), flashlight battery core, aluminum and iron piece with holes from old hair curler, wax T-bar. Above, heavy aluminum with high silicon, charcoal, deer femur with cut end. These items together had probably been collected by pack rats.*

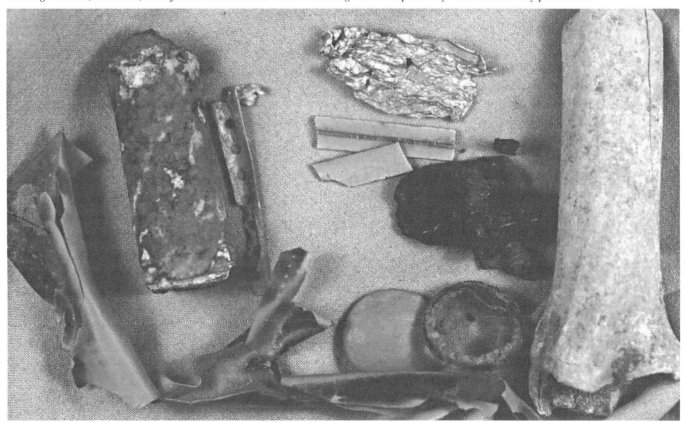

it was chipped or broken and eventually found its way into the arroyo with other debris. I believe the site may have been visited occasionally from the late 1940s into the 1960s, and that the rain jacket, the cross, and the milk bottle may be associated. It is possible that the milk bottle held wild flowers possibly associated with the presence of a woman or child at the site one or more times. There were also the other wide-mouth bottle pieces found under the cross, which would indicate to me more than one visit there. In the ten years we visited the site, six of them were in May, July, August, and September. Many wild flowers grew nearby.

After we discovered the initial UFO touchdown area in the upper arroyo, it could be seen that some monitoring of the site after the crash would be desirable until the natural foliage, grasses, and shrubbery returned. It would make sense in cooler weather to have a small fire with drinks or snacks there. We also excavated from the pit what we believed to be some WWII military K-ration can rings. We suspect that military individuals came initially to check on the upper and lower arroyo gap sites, and trusted civilians possibly came later. Even 60 years later, foil shards like those analyzed by Steve Colbern (Chapter 4) still appeared on the surface after a rain.

At one time I think the cross was standing with the milk bottle and jacket on the ground. Underneath, over the years, as the prairie dog complex expanded, the mound grew outward pushing over the cross and burying the bottle or bottles (which eventually broke) with the red jacket just under the surface. Combining the age of the lichen on the cross, its placement, the weathering of the cross notch, and the wide-mouth bottle speculation, I believe the cross was moved to the gap from another location in the late 1950s to the early '60s.

The color red may be significant here. When we started inspecting the small shoe sole. (See Chapter 13.) It was apparent that some faded red material (Figures 13.2 and 13.3) made up the outside covering of the shoe sole. There was much evidence of visitors at various times at the site after the crash. Might the red jacket, spread out under the cross and the little shoe sole be connected? Was there a memorial of some kind?… We wonder.

Was it plausible that all of the material found under the cross and in the pit may have been brought in by high water in the arroyo? Several factors need to be considered here. The braids of the arroyo act as channels for a small delta. When one channel clogs up, water finds its way around the choked-up area into a new channel or more likely, comes back into an older channel. Thus a four-foot depth of water in a deep arroyo channel a mile upstream dissipates into the 200-yard-wide area of the alluvial fan. Water in the main arroyo channel might not be over two to three inches deep when it rains. Ninety percent of the water spreads out in the arroyo and is absorbed by the sandy soil or puddles until it evaporates. There was no lichen on the back side of the cross next to the ground. This pretty well eliminates the possibility that the cross and the artifacts found under or near were brought in by running water in the last forty years or so. The main, active water channel now runs about 60 yards away.

There are other factors to consider. The cross and pit area is uphill from the nearest arroyo channel. There are no signs of running water near this area. If the charcoal had been brought in by water, it would not have been concentrated in the pit. Twenty-one soil samples were dug in the gap nearby and no similar material was noted. I think it is more than happenstance that the cross, the earth-made red jacket, the milk bottle and other glass, the two pieces of metal, the T-bar charcoal, and other items were there together. For whatever reasons we can only speculate.

The cross and pit site was near the gap. Why anyone would build a fire at this site is a mystery, unless it was for warmth. But there are lots of various types of shelter on the sides of the arroyo, that would provide some protection, especially from the Plains winds. The cross location certainly was not a camp site. With the dispersion of the charcoal in the pit, I feel that there was more than one fire, at different times, as mentioned before. While excavating at this site in 2004, we discovered a large fire ring of placed stones in a sandy area, about 20 yards from the pit. It was evident that one end of the cross had lain on a fire, and I also believe at a later time when the cross was made, it became a handy crosspiece. The pit accounted for two of the fifteen kinds of unusual foil found at the site, as well as a number of other interesting items. Again water was not a factor, as the pit was up hill from any arroyo braids. The concentration of the material under the old prairie dog mound, suggests rodent activity. The Banner Tailed Kangaroo Rat (also known as a pack rat) lives in the area and collects objects to put in its nest or burrow. We cut through several tunnels leading to these burrows in our excavation.

## CHAPTER 3
### REFERENCES

1    Campbell, Art; Personal correspondence with Bureau of Land Management, Botanist and Glass Manufacturers.

2    Toulouse, Julian Harrison; *The Thatcher Glass Story*; Thatcher Manufacturing Co.; p. 17-20.

3    Toulouse, Julian Harrison; *Bottle Makers and their marks*; Thomas Nelson, Inc., Camden, NY; 1971; p. 496.

Several other very good references were very helpful. They include:

William Steinman's *UFO Crash at Aztec, A Well-kept Secret.*

*The Dictionary of Economic Plants*, published by Stechert–Haffner, Service Agency, NY.

# CHAPTER 4

## THE FOIL SHARDS "MAY NOT HAVE ORIGINATED ON EARTH"

### OVERVIEW

This chapter could be a book unto itself. It is a very thorough analysis of metal foil from a sandy arroyo in New Mexico and how, in these pages, it implies a space ship coming to Earth in July 1947. We know from many sources, including Mars meteors, moon rocks, and comet debris, that isotopic ratios and molecular content of metals are much different on Earth than other bodies from space. Five pieces of the shard metal foil were analyzed, and one heavy piece of metal. With the help of very specialized lab equipment and excellent scientific work, only three isotopes, antimony (Sb), copper (Cu), and nickel (Ni) of the W-1 sample, were stable enough to perform isotope abundance calculations from ICP-MS data (Table 2, page 22). Each metal shard had up to 56 separate identifiable elements in a coating. Many elements were probably already in the metal, but at least a dozen were obviously added. The technology to produce these highly sophisticated coatings was not known in 1947, and all quite probably did not originate on Earth.

*Figure 4.2. **Chuck Wade with wife Nancy** at the Plains of San Augustin crash site in 2004. Chuck orchestrated a great, productive dig. He designed four collapsible tripods for screens and arranged for Navajo workers who screened 28 excavations, each 40 inches across by 12 to 16 inches deep (1 m x 30-40 cm). Chuck, from Gallup, NM, is the retired owner of Wade Building and was raised in Corona, NM, a few miles north of the Roswell Site.*

The metal foil was curious to look at when we found it. I saw some of it sticking out of the arroyo soil after several hard rains. The pieces seldom exceeded a few inches square. Chuck Wade, an associate, had one about 5 inches (13 cm) long and 3 $^1/_2$ inches (9 cm) wide, after it was flattened out. A lot of the metal pieces found were crinkled with ridges and valleys that ran more or less parallel to each other, similar to how a thrust-faulted mountain chain, like the Appalachian Mountains, appears today (Figure 4.3).

We contacted the Aluminum Association, which sent its worldwide patent formulas for all manufactured aluminum

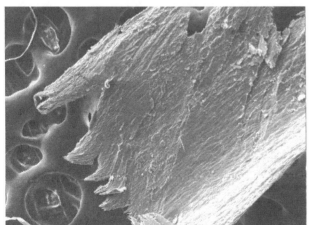

*Figure 4.1. **Columnar metallic crystals** from W-6 metal phase. SEM magnification at 85x. Metal phase refers to the reverse side of metal that is opposite the coating.*

### Table 1

**COMPOSITE METAL COATING ELEMENTS IN SIX CRASH SITE METAL ARTIFACTS**
**In Order of Abundance**

| | | |
|---|---|---|
| **Aluminum** | Boron | Beryllium |
| **Iron** | Cerium | Samarium |
| **Silicon** | Tin | Bismuth |
| **Calcium** | Cobalt | Ytterbium |
| **Manganese** | Rubidium | Erbium |
| **Magnesium** | Molybdenum | Iodine |
| **Potassium** | Lanthanum | Cadmium |
| **Titanium** | Yttrium | Cesium |
| **Copper** | Neodymium | Germanium |
| **Phosphorus** | Lithium | Silver |
| **Zinc** | Thorium | Antimony |
| **Sodium** | Uranium | Thallium |
| Vanadium | Hafnium | Holmium |
| Chromium | Arsenic | Europium |
| Barium | Tungsten | Selenium |
| Nickel | Niobium | Thulium |
| Gallium | Gadolinium | Platinum |
| Strontium | Gold | Mercury |
| Zirconium | Praseodymium | |
| Lead | Dysprosium | |

*Significant coatings on metal in bold above. Others are trace elements that probably were in the metals originally.*

products. There were about a dozen formulas for foil. The commonly used formula is for household foil made by Reynolds Metals and Alcoa. Their basic formula was aluminum (99.45%), silicon (0.50%), and iron (0.5%). Standard kitchen foil has these three elements.[1] Metal foil shards W-1 to W-6 found in the arroyo had 30 to 56 elements in a single piece.

One of the first investigations of the foil was to see if it could have come from domestic commercial sources in the area. We found that aluminum foil was not sold in the west-central New Mexico area until about 1955. It became popular when home freezers came into vogue in the mid-'50s. It was used as an early meat wrap to prevent freezer burn. Although most ranchers have freezers today, they could not be used until a

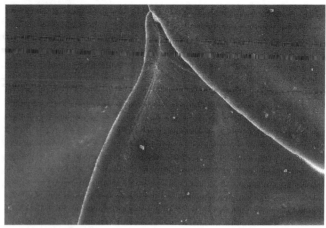

*Figure 4.4.* **Folded over seam, upper center of W-1.** *Some metal shards had original edges; most were torn,. This edge was folded under and crimped. SEM image 100x. Seam width is about 3 human hairs wide, or 210 microns. Debbie Ziegelmeyer found a similar piece at a Roswell site in 2002.*

as we know it today, was a byproduct of surplus WWII aircraft. Both Chuck Wade and I found at least one bit of foil that looked like it had been folded and tucked into the soft sand. Chuck's looked like a couple of small taco shells folded together. This gave weight to a hypothesis and strong evidence that the site had been monitored, and this foil had been picked up, folded, and placed back under the sand to get it out of sight. This may have occurred at the initial military cleanup or during monitoring activities later. We were able to isolate a piece of foil that had an original edge. We believe the foil may have originally come in shingle-like pieces. When we found straight edges (not torn), the edge was folded under as seen when magnified by the SEM, shown in Figure 4.4 above.

*Figure 4.3.* **Hills and valleys on W-3.** *Looks like a topo map of West Virginia's Appalachian Mountains. Foil metal sheets were probably flat whey the came into the arroyo. It is possible they were crinkled in the UFO. Pressure in the alluvial fan over years we believe pushed in ridges. W-3 had a total of 44 elements in its coating.*

stable source of electricity came to the Plains area about 1953–54. A ranch wife who lived near the Arroyo crash site said she did not remember foil on a roll for consumer use until around 1956 when her family got their first barbecue grill. Rural ranchers had other time-honored ways of preserving and preparing food for storage, including smoking their meat and canning their fruits and vegetables.

The first known use of aluminum foil in the U.S. was in 1914 for candy and gum inner wrappings, and later for cigarette packaging. Some sheet foil was used in the 1930s in commercial kitchens of large restaurants and hospitals. Wide use of it was limited during the Depression years due to cost. Aluminum foil,

*Figure 4.5.* **Folded metal foil shard as found in 2004.** *Natural movements in arroyo sand could not cause this. It had been picked up, folded, and stuck back into the sand. Other evidence indicated the crash site was monitored for about 20 years.*

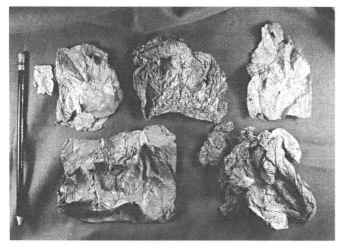

*Figure. 4.6. **Foil found 1996-2005**. Most of this foil was found protruding from arroyo sand after a rain or high wind. Foil was extremely brittle and varied in thickness. More foil was found under the surface of sandy loess soil.*

I had been working the San Augustin site since 1995, when I started picking strange things off the ground. As a Western historian with three books published and an interest in Western artifacts, I was in somewhat of a position to know what should be on the ground and what should not. The materials analyzed in this book all came from the San Augustin site, including the strange metal pieces or foil, which will be expanded upon in this chapter.

Chuck came along at the right time, as I had pretty well exhausted the surface finds in the immediate crash site vicinity, and had expended considerable resources traveling to and from the site nearly every year. I had become familiar with the local people and had a good relationship with the property owner. Chuck offered to bring some labor to the site to do considerable excavations the next summer. We agreed to meet at the Eagle Guest Ranch in Datil the following June. A man of his word, Chuck was there with several two-man crews of Navajo workers and four collapsible tripod frames to hold the sifting screens he had made. On the way

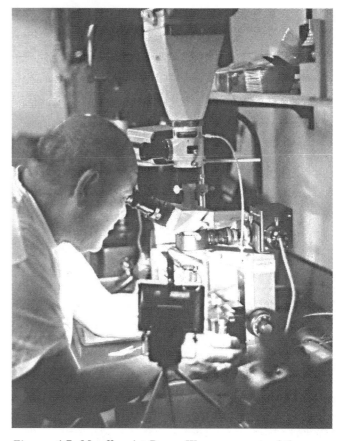

*Figure. 4.7. **Metallurgist Bruce Wong** was one of the many scientists who worked on this project. Wong analyzed the foil in 1997 and 1998. This lab used a Kevex 7500 Energy Dispersive X-ray for analysis. This scientist said the foil was so unusual that "it must have been scraped from an aluminum plant floor." This was before the ICP-MS process was used.*

**LABORATORY EQUIPMENT**
- Six samples of the metal were tested at the laboratory. Images of the samples were taken using an 8X–40X lighted dissecting microscope.
- Samples were again imaged using a higher power microscope of higher magnification (100X–400X).
- SEM flakes of each sample were removed and mounted on aluminum posts for scanning electron microscopy. The scanning electron microscope directs and focuses a beam of high energy electrons on a sample in a vacuum chamber. When the electrons are reflected by the sample, they form a high resolution image on a monitor.
- Energy Dispersive X-ray (EDX) chemical analysis is done in the SEM. The EDX provides chemical analysis of minute particles using the SEM's high energy electronics beam.
- Inductively Coupled Plasma Mass Spectrometry (ICP-MS) is a type of mass spectrometry that is highly sensitive and capable of simultaneous analysis of a range of metals at the parts per billion level. It is based on bringing together an inductively coupled plasma as an ion source with a mass spectrometer as a method of detecting the ions.
  1) ICP (Inductively Coupled Plasma). This is a high-temperature plasma (ionized argon) sustained with a magnetic field, which acts to produce ions. The plasma is approximately 6,000°C. It is three times hotter than a welding torch and is equal to the temperature of the surface of the sun.
  2) MS (Mass Spectrometry). The ions from the plasma are extracted through a series of cones into a mass spectrometer, usually a quadrupole. The ions are separated on the basis of the mass-to-charge ratio, and a detector receives an ion signal proportional to the concentration. The concentration of a sample can be determined through calibration with elemental standards.
- A Neodymium-Iron-Boron (NIB) magnet was also used to determine whether any of the metal was ferromagnetic. A pendulum with a lead weight was passed over the samples for gravitational or magnetic fields.

*Figure 4.8. **Author Art Campbell** in first excavation, 1997. Screen was borrowed from a rancher. The old prairie dog complex was easy digging, but one needed to watch out for rodents, which carried deadly hantavirus, and also for occasional snakes.*

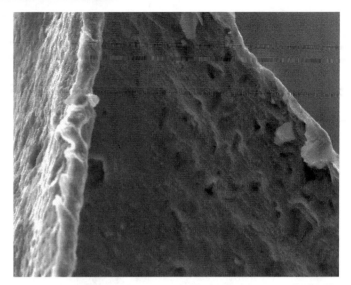

*Figure 4.9. **Scanning electron microscope image** of badly weathered shard of W-4. Metal pieces found closer to the surface showed more weathering. Those that were excavated later were in much better condition. SEM scan 100x. Coating contained 49 elements.*

down to the site from Gallup, they bought a case of honeydew melons and some tomatoes. By the time our desert camp was all set up, it reminded me of a Mexican fruit market, until a 60 mph wind rearranged things a little. We were there three to four days and dug 28 test holes, where Chuck's Navajo crew found most of his metal foil. I had an equal amount I had found on the surface in the previous nine years I visited the site. What he found was his, and he was thrilled. We met the next year in 2005, but it was not nearly as productive as the 2004 dig had been. I was called home early when my wife sustained a serious hip injury.

Chuck and I had no formal arrangement about what to do next or how to proceed. We each just sort of did our own thing. I took or shipped my material to various labs and went on working on this book. Between 2005 and 2010, Chuck developed a great website and started to expand his contacts as his project grew. After we had done a few radio shows and been to a few conferences, many amateur scientists wanted to analyze our materials but lacked professional lab equipment and expertise.

We both had problems finding credible professional people who would also publicly stand behind their work. Later in the project we were able to find scientists who would give us scientific data with valid testing that would zero in on molecular and isotopic analysis. Data from these tests indicated whether our material originated on earth or elsewhere. One such report begins on page 23. Sending it out to professional labs was also problematic. One issue was that many labs would not take UFO-related material if they knew of our research. We could get around this by not mentioning UFOs, but this had its drawbacks. Hardly anyone who was qualified to do sophisticated testing wanted their name or organization mentioned. We needed to publish or at least speak about these organizations to add credibility to our project. Chuck was relentless in seeking the right people, and his persistence paid off, as can be seen by this work.

By the fall of 2006, Chuck started going to UFO conferences and did some speaking. Chuck and his metal were becoming very well-known. Later, through his excellent website, he was able to tap scientific energies through amateur scientists who had been interested and working in and around the UFO and free energy fields of interest. Sometime in 2009 or 2010, Chuck made contact with John Rao, who had initially acquired the rights to the UFO Congress. John offered to test the five pieces of metal

*Figure. 4.10. **Various materials** came up in screen which is 1/4 inch wire mesh. Dirt was first scraped from pit, layer by layer, and swept into dustpan for screening. Above shows a chunk of foil and what may be a rusted K-ration can ring.*

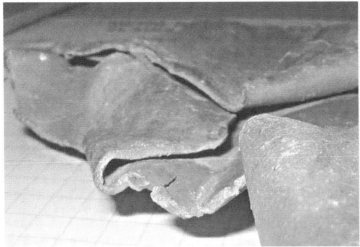

*Figure 4.11.* **Heavy piece of metal** *(W-6) from crashed craft. Found on the surface 200 feet from the gap/crash path by one of our 2004 Navajo crew. This may have been part of the upper exterior of the craft. The absence of crucial brown dots indicate that the upper part of the craft may have used a different power system.*

*Figure 4.12 (upper right).* **Freshly cut piece of W-6.** *This was the thickest of the metal shards found at the site, about 1mm thick. Our scientist had difficulty getting a clean, uncontaminated sample to analyze. SEM image 85x.*

*Figure 4.13. (right)* **Coating on W-6 peeling off.** *Outlined area in the image on the upper right is the location of this close-up. Coatings were adhered to base metals. This is the first isolated coating we found peeling from base metal. The coating is primarily aluminum, with other major and minor elements. These included iron, magnesium, mercury, chromium, zinc, and silicon.*

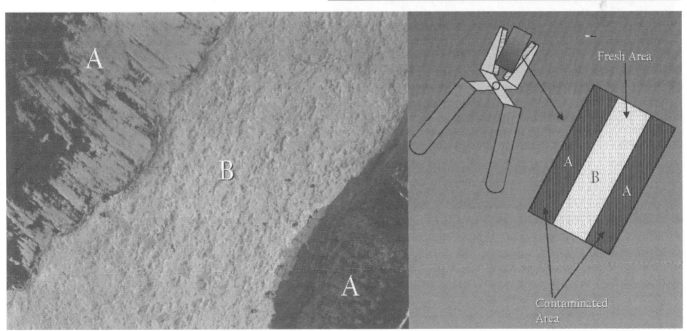

*Figure.4.14.* **W-6 freshly cleaved area.** *The sides of the pliers went into material at locations A, partially contaminating the aluminum. The sample then cleaved open in the middle without direct contact, leaving a clean area at location B for sampling.*

foil shards in addition to one other heavier piece. The material was sent to Steve Colbern, a southern California chemist materials-analyst

Steve Colbern's analysis of the six metal samples and his findings are truly remarkable. His work picks up the reins from several scientists who preceded him, laying important groundwork. There were at least five labs that did serious work before Steve's 2010 analysis. The SEM magnifications (dark photos) that appear in this chapter were done by another lab scientist, who is paralleling Colbern's work. Early on, the metal samples were dispersed to different laboratories. Being in different parts of the country worked well for us. We had initial concerns that what we were finding was very relevant to something that once had been extremely secret. As more important material was found, for security reasons, having materials at different locations turns out to have been a good idea.

## THE COLBERN REPORT

### STEVE COLBERN'S BACKGROUND

Steve has over 20 years experience in chemical materials science. He is well known in his field and a graduate of UCLA. He is an eminent chemist at applied science with a specialization in materials analysis. He holds some patents for his work and discoveries. One notable achievement was on Synthesis of Ultra-Pure, Sol-Gel Derived, Silica Glass for Ultraviolet (UV) Applications. Another aspect of some recent work in materials science involves carbon nanotubes (CNT), which are extremely strong and have superior electrical conductivity. He is a recognized expert in the field of Nanotechnology.

One of Steve's roles in the analysis of the unusual metal and other materials found at the crash site, was to find labs to do the initial ICP-MS and Isotope Analysis. Reports from several labs were synthesized by him for this book. This work was coordinated from labs on the West Coast, several in the Midwest, and one in the Eastern U.S. Their findings are chronicled in this book.

Some of these efforts involved electrochemistry, the use of scanning electron microscopy, chem informatics, and other sophisticated scientific processes and equipment. Steve's goals for his research are to "use different interdisciplinary approaches to find breakthrough and novel solutions to scientific and industrial problems." Steve is currently a member of Technical II at YTC America. Steve's experience in chemistry, materials analysis, and alternative energy has given him a good reputation in related fields of science and engineering. We were indeed fortunate to have had Steve's Colbern's expertise and research skills in this analysis.

### ISOTOPIC RATIOS

Some of Steve's important work here primarily deals with isotopic ratios. The best explanation we have found was written by Steve about isotopes naturally occurring on Earth and on other planets. It is as follows:

**It is well known that all matter is composed of atoms. And atoms consist of a nucleus surrounded by an electron cloud. All nuclei, other than the simplest hydrogen, are made up of both protons and neutrons.**

**Atoms, which have the same number of protons are all the same element, like aluminum or carbon. When the number of protons between two or more atoms is the same but the number of neutrons is different, these atoms are called isotopes. For example, one isotope of carbon has six protons and six neutrons and is called carbon-12. Another isotope has six protons and seven neutrons and is carbon-13. Naturally occurring on the Earth, carbon is a mixture of 98.9% carbon-12 and 1.1% carbon-13. This will be true for all of the naturally occurring terrestrial carbon. If a sample [of] carbon was found to be a 50% carbon-12 and 50% carbon-13 mixture, we would have to conclude that the sample was not naturally occurring on the Earth.[2]**

### TABLE 2 – ISOTOPIC RATIOS OF SUITABLE ELEMENTS IN SAMPLE W-1

| ELEMENT | ISOTOPE | SAMPLE ISOTOPIC ABUNDANCE (%)* | TERRESTRIAL ISOTOPIC ABUNDANCE (%)‡ |
|---|---|---|---|
| Antimony | $Sb^{121}$ | 49.58 | 57.36 |
|  | $Sb^{123}$ | 50.42 | 42.64 |
| Copper | $Cu^{63}$ | 48.84 | 69.15 |
|  | $Cu^{65}$ | 51.16 | 30.85 |
| Nickel | $Ni^{58}$ | 35.31 | 68.08 |
|  | $Ni^{60}$ | 32.41 | 26.23 |
|  | $Ni^{61}$ | ND | 1.14 |
|  | $Ni^{62}$ | 32.28 | 3.63 |

**ISOTOPIC ANALYSIS OF SAMPLE W-1**

Antimony (Sb), copper (Cu) and nickel (Ni) were the only elements present in the samples which were suitable to perform isotopic abundance calculations on from the raw ICP-MS data.

These elements were suitable for this analysis because there are no analytical interferences with their isotopes from other isotopes found in the samples.

The results of the isotopic abundance calculations for sample W-1 are shown in the table. These results are very unusual and show extremely skewed isotopic ratios in the three tested elements, relative to the normal terrestrial amounts of the isotopes in each of these elements.

\* **May not be from Earth**
‡ **From Earth**

*The results are conclusive after two tests with antimony (Sb, two with copper (Cu), and four with nickel (Ni). These elements from W-1 seen in Sample Isotopic Abundance column above probably could not have originated from Earth.*

## THE REPORT
Steve's 42-page report on the metal samples analysis from the 1947 UFO crash on the Plains of San Augustin was detailed and comprehensive. Excerpts and highlights from the report are included below:

## BACKGROUND INFORMATION
Six metal samples were obtained from Mr. Chuck Wade, who stated that he and a digging crew excavated the samples in June 2004 from the desert floor on the plains of San Augustin, New Mexico. This area was reportedly the site of the July 2, 1947, crash of a small, extraterrestrial craft.

Some of Mr. Wade's materials have been analyzed previously by light and scanning electron microscopy. No other analytical results from these materials have been published to date.

## ANALYTICAL PROCEDURE
Six metal samples were given to [the lab] for analysis. Digital images of the samples were taken, using a dissecting microscope, at 8X–40X magnification. The samples were then imaged using another light microscope, capable of much higher magnification (100X–400X).

Flakes of each sample were then removed by cutting with a surgical scalpel and mounted on aluminum posts for scanning electron microscopy (SEM) imaging and energy dispersive X-ray (EDX) elemental analysis, to determine the presence and distribution of elements in each sample.

SEM magnifications from <100X–15,000X were employed. EDX area scan, elemental mapping, and point-and-shoot analyses were also employed.

Small pieces of each sample (~10 mg) each were then cut off, dissolved in nitric acid, and analyzed by inductively coupled plasma mass spectrometry (ICP-MS), to determine the concentrations of the major component elements in each sample, along with the trace element abundances. The ICP-MS raw data was then used to determine the relative abundances of isotopes of three elements in one of the samples.

The samples were also exposed to the field of a Neodymium-Iron-Boron (NIB) magnet to determine whether they are ferromagnetic. A pendulum with a small lead weight attached was also passed over the samples as a simple test for gravitational, or magnetic, fields emitted from the samples.

## COLBERN REPORT — THE SAMPLES
### APPEARANCE & PHYSICAL CHARACTERISTICS
The samples (W-1 – W-5) were all shards of silvery sheet metal, which resembled sheet aluminum. A sixth sample, known as W-6, was a collapsed piece of much heavier folded aluminum. *[See Figure 4.11]*. Two of the samples had a tan, or greenish-tan, outer coating which appeared to be a protective layer.

All of the samples, with the exceptions of W-2 and W-6, had many ridges in the material, and had a crumpled appearance.

All of the samples were able to be bent by hand, with sample W-6 being the only exception. This sample was thicker than the others, and its increased thickness may have accounted for its greater strength.

## LIGHT MICROSCOPY
All samples appear to be a sheet metal, which resembles aluminum. All of the samples are very thin (~0.05 mm), except for sample #6, which is approximately 1.0 mm thick. Samples #2, #3, and #6 had thin surface coatings, which appeared to be a few microns in thickness.

These coatings later proved to be of different composition than the base metals. These coatings could be removed by scraping with a knife or other metallic instrument but were bonded fairly well to the metal. Samples W-2 and W-6 appeared to have the thickest coatings. The coating on sample W-2 was tan in color, the coating on sample W-3 was brown, and the coating on sample W-6 was greenish.

The metallic portions of samples W-1 and W-3 appeared to have round structures, or crystals, embedded in the metal, which were several microns in average diameter. Sample W-6 had a layered structure composed of what appeared to be columnar metallic crystals. *[See Figure 4.1]*. The metallic portions of the remaining samples appeared to be very uniform, under light microscopy. *[Metal portions, or metal phase refers to the base metal under the surface or coating.]*

## SEM IMAGING
Scanning electron microscope (SEM) images were taken of all of the samples, at magnification ranging from 25X to 15,000X. The SEM images showed ceramic-like crystals, cracks, and pits in the outer coatings of the coated samples, and metallic crystals in the metallic portions of the samples, especially sample W-6.

## EDX DATA
Energy Dispersive X-ray (EDX) elemental analysis was performed on all six Wade samples, in the course of obtaining the SEM images. This analysis enables detection of elements present in relatively high amounts, and gives the relative proportions of each.

EDX elemental mapping and point-and-shoot were done on selected sample areas, in addition to the standard EDX spectra, which are elemental abundance averages over the area imaged. EDX mapping shows a map of the relative concentrations of the elements detected in the imaged area, while EDX point-and-shoot displays EDX spectra at selected points of an imaged area, highlighting differences in the composition of imaged features.

The major component of the metallic portions of all of the samples proved to be aluminum (Al). All the samples appeared to be composed of aluminum alloys, with varying amounts of alloying elements. Other elements detected included beryllium (Be), carbon (C), oxygen (O), sodium (Na), magnesium (Mg), silicon (Si), phosphorus (P), sulfur (S), chlorine (Cl), potassium (K), calcium (Ca), titanium (Ti), iron (Fe), and palladium (Pd).

The coating layers of the coated samples were much different in composition from the uncoated portions of the samples. Aluminum was still a major component of the coatings but was present to a lesser degree than in the metallic portions of the samples.

The amount of oxygen in the coatings was much greater than in the metallic phase of the samples, indicating that the aluminum was probably present as an oxide layer, rather than as free metal. The proportions of carbon, silicon, and chlorine in the coatings were also higher than in the metal, indicating the probable presence of metallic silicates, carbonates, and chlorides as components of the coatings.

All the elements detected in the metallic phases were also present in the coatings. Some elements were also present in the coatings which were not detected in the metallic phases; these included nickel (Ni) and barium (Ba). The coatings of samples W-1 and W-6 were also quite similar to one another. The coatings on samples W-2 and W-3 had not been analyzed by SEM/EDX as of the date of this report, but may be in the near future.

The EDX mapping of the coating of sample W-1 indicated that the oxygen, silicon, potassium, calcium, and carbon in the coating tend to be concentrated in particles on the surface, which appear lighter in the SEM images. The EDX point-and-shoot technique confirmed this. The majority of the darker coating surface appears to consist of aluminum oxide.

### MAGNETIC AND ELECTRICAL ANALYSIS

None of the samples were attracted to a strong Neodymium-Iron-Boron magnet and are therefore not ferromagnetic. All of the samples were tested with a volt-ohm meter and were found to conduct electricity. Quantitative values of sample resistivities will be determined when more of each sample is available.

### RAMAN SPECTROSCOPY

Raman spectroscopy at 532 nm laser wavelength was carried out on the samples primarily to test for the presence of carbon nanotubes, as this is a sensitive and reliable test for their presence in a material.

The presence of carbon nanotubes in the samples was suspected because of the previous detection of carbon nanotubes in an alien implant sample, which was recently (2008) removed from the body of an American materials scientist. Carbon nanotubes are currently being actively researched in Earthly materials science because of their uniquely high strength-weight ratio, and electronic properties, and it was hypothesized that much of the alien technology may utilize these materials.

The Raman evidence for the presence of carbon nanotubes is inconclusive for samples W-2 through W-6. Weak peaks appear in the area of 1200 cm-1 to 1600 cm-1, which could be produced by single-walled carbon nanotube D and G bands, but the signals are too weak to permit positive identification as such. Sample W-1, however, does appear to have peaks that have a much higher probability of being caused by the presence of single-walled carbon nanotube D.

### ICP-MS ANALYSIS

Pieces of all of the samples were subjected to trace element analysis by Inductively Coupled Plasma Mass Spectroscopic (ICP-MS) analysis, performed by BodyCote testing lab, in Santa Fe Springs, CA.

This analysis involves dissolving small amounts of each sample (10 mg) in a mixture of nitric and hydrochloric acids, and passing the solution through a plasma torch. The resulting plasma, containing ions (charged atoms) of the elements in the sample is then passed through a mass spectrometer, which sorts the ions by charge/mass ratio.

This analysis provides sensitive and quantitative results on the amounts of all elements present in the sample. Amounts of most elements below parts-per-million (ppm) levels can be detected using this analysis.

Sixty-eight (68) elements were tested for, with fifty-six (56) elements being detected in at least one sample. Aluminum (Al, average concentration 95.6%) was the most abundant element in all of the samples, followed by iron (Fe, ave. 2.10%), silicon (Si, ave. 1.02%), calcium (Ca, ave. 0.41%), manganese (Mn, ave. 0.25%), magnesium (Mg, ave. 0.22%), potassium (K, ave. 0.07%; 707 ppm), titanium (Ti, ave. 0.03%; 333 ppm), copper (Cu, ave. 0.03%; 317 ppm), phosphorus (P, ave. 0.03%; 310 ppm), zinc (Zn, ave. 0.03%, 309 ppm), sodium (Na, 0.02; 155 ppm).

The amounts of the most abundant elements in the six samples varied widely with the specific sample, implying differences in function.

### OTHER TESTS PERFORMED

Samples W-1 through W-6 were placed on a flat surface, and a pendulum, constructed from a 4 oz lead weight tied to an 18" long piece of monofilament nylon line was passed over the samples. When the weight passed over the samples at close range (< 2") the weight consistently showed a noticeable deflection away from the sample.

These results are similar to those obtained from two similar tests done on all six samples by Chuck Wade at the 2010 UFO Congress, in Laughlin, NV.

## COLBERN REPORT — DISCUSSION

### APPEARANCE & PHYSICAL CHARACTERISTICS OF SHARD SAMPLES & W-6

The samples are composed of aluminum alloy sheet, some of which are coated with what appears to be a protective coating. The samples had some soil attached when first received and had clearly been buried at one time.

The corrugations (wrinkled surfaces) on W-2, W-3, W-4, and W-5 are reminiscent of the type of bending which can occur from sudden shock, as in an aircraft crash, although it cannot be ruled out that the samples could have been manufactured in this form.

The samples were composed aluminum alloys, all having a low content of copper, and with unusual alloying/trace elements, many of which were unheard of as components of aluminum alloys in 1947, and they are unlikely to have been introduced during the aluminum manufacturing process in that era.

These facts are consistent with the material being debris from the crash of an aircraft or spacecraft at the San Augustine desert location. If the crash did occur in 1947, the material seems inconsistent with the materials that were commercially available at that time, and are possibly too advanced to have been produced by the technology of that time period.

The mechanical strength of the materials is not extraordinary, however, and seems well within the normal limits of the strength of commercially available aluminum alloys. The materials could all be bent, torn, and cut with relative ease.

It is not known where these samples came from in the structure of the craft, however, and it is possible that they came from interior structures, which did not require extreme mechanical strength. If this is the case, then samples from the exterior of the craft may show much more mechanical strength and toughness.

The layer of ceramic-like material, seen on some of the samples (W-2, W-3, and W-6) under light microscopy, is interesting, and appears to be some type of protective layer placed over the metal. This type of technology was probably not available in 1947 *[on Earth]*.

One of the materials (sample W-6) also appeared to have a layered structure, which is not typical of commercial aluminum alloys. *[See Figure 4.14]*. The SEM images of the materials also show surface coatings on the samples, which appear to be applied, and are not the simple aluminum oxide surface layer which forms naturally on standard aluminum alloys. These coatings have pits, and pores of somewhat regular composition. There are also particles on the surface, which EDX indicated have different composition from the remainder of the coating.

The layered structure of sample W-6 is also very apparent in the SEM images. This type of structure is not seen in aluminum alloys, and is more reminiscent of the structure seen in some titanium alloys, or a more complex material, applied in layers by chemical vapor deposition or some similar technique.

The EDX area data confirmed that there is a coating on some of the samples, which differs significantly in composition from that of the metallic phase *[base metal the coatings were attached to]* and contains many elements not found in the metal.

EDX mapping and point-and-shoot data confirmed that the coatings are not homogeneous and contain particles with different elemental composition than the rest of the coating. These coating particles contain increased amounts of oxygen, silicon, potassium, calcium, and carbon.

The large array of elements detected in the samples by the ICP-MS testing is an indication that these are very complex aluminum alloys, which contain unusual alloying elements, and are unlike typical aircraft aluminum alloys that were available in 1947.

The presence of relatively large amounts of iron, calcium, silicon, zinc, relative to what is usually present in aircraft aluminum alloy, appears to indicate that the alloys may have been intended for an application requiring good electrical conductivity, as these elements do not decrease the electrical conductivity of aluminum as much as most other common alloying elements.

The rare earth metals may have been added to the alloy to strengthen the material, as a large amount of research has been done in recent years on the use of these elements to strengthen aluminum alloys.

A dedicated isotopic analysis should be done on this sample to confirm this conclusion.

The Raman data, indicating the possible presence of carbon nanotubes in sample W-1, was very intriguing. These materials were discovered in 1991, have unique mechanical and electrical properties, and are currently an active area of investigation in Materials Science. Carbon nanotubes are being investigated to strengthen metals and create embedded electronic components.

These potential uses for carbon nanotubes result raises the possibility of the samples being advanced "smart metal" materials, containing carbon nanotube electronics. The results of the pendulum test indicate that the tested samples may be emitting an electromagnetic, or gravitational field, which supports this hypothesis. Gaussmeter testing should be done on these samples to investigate whether a magnetic field is present.

## COLBERN REPORT — CONCLUSIONS

1) These samples contain very unusual alloying elements which were **not present in aluminum alloys in 1947**. If these samples are from an aircraft which crashed in that year, they are very unusual on that basis.
2) The coatings on the samples are also unusual because conformal coatings of this type, which are blended with the metal, and rich in silica, titania, magnesia, sulfate, phosphate, and chloride, were almost **certainly not available in 1947**. The coatings on the samples are also somewhat similar to coatings on implants removed from people claiming alien contact.
3) The carbon nanotube indications observed in the Raman spectra of the samples indicates the possibility that the samples may be "smart metal" materials, which contain carbon nanotubes as electronic components, or to strengthen the materials. Since the mechanical strength of these samples was not unusual, they should be tested for unusual electrical characteristics.
4) The isotopic ratios of three elements in sample W-1 (antimony, copper, and nickel) were extremely skewed with respect to the terrestrial ratios for these elements, and there is, therefore, a high probability that **the samples came from an extraterrestrial source.** These extremely skewed isotopic results are again reminiscent of those obtained from alleged alien implants, and from an alleged piece of the Roswell crash debris which was analyzed by the late Dr. Russell Vernon Clark.

*[Editor's note: In this instance, "skewed" refers to metal formulations not known on earth.]*

5) The results of the pendulum test indicate that samples W-1 and W-6 may still be emitting gravitational or magnetic energy, which greatly increases the probability these samples are **nanotechnological "smart metals" and of probable alien origin** as well.
6) Further microscopic testing should be done on these materials to determine their internal structures. More testing should also be done to determine the existence, extent, and profile of any gravitational, magnetic, or electric fields the samples may be emitting, and their source of energy.

The results of the isotopic analysis of the ICP-MS results from sample W-1 (Table 1) indicate that the isotopic abundances of each of the elements tested (antimony (Sb), copper (Cu), and nickel (Ni)) **differed significantly** from the isotopic composition of the same elements derived **from terrestrial sources.**

For elements heavier than boron, differences in isotopic composition of more than approximately 1% from the usual terrestrial isotopic abundance pattern, indicates a high probability that the **material originated from a non-terrestrial source.**

All of the elements tested differed from the terrestrial abundances by much more than this. These elements in sample W-1 therefore **may not have originated on Earth.**[3]

*[Emphases mine]*.

## CUTTING EDGE EQUIPMENT — TIMELINES

One cannot really appreciate the extraordinary work of the material scientists like Steve Colbern and his associates until we see a total, comprehensive report. (There were three scientists working on this project at one time). Taking small samples and making a diagnosis from flat shingle-like pieces of metal (with up to 56 known elements) is like finding a piece of a tire rim, a cup of brake fluid and a fragment of windshield wiper and concluding a moving automobile. None of this would be possible without sophisticated equipment, which is new and still developing. The following timelines are important milestones to this groundbreaking work:

**1590.** Two Dutch eyeglass makers, Zaccharias Janssen and his son Hans, experimented with multiple lenses placed in a tube. This is the forerunner of today's **compound microscope.**

**1939.** An extraordinary letter from scientist Albert Einstein to President Roosevelt on the eve of WWII acts as an accelerant to the development of this equipment. Roosevelt listened and later wrote, "This requires action" when Einstein discussed the implications of a nuclear chain reaction and powerful explosives later known as "atomic bombs."

**1940.** In May of this year, at the beginning of WWII, promising results at Columbia University led scientists to propose **isotope research** while studying uranium isotope separation.

**1947.** It is believed the Plains of San Augustin **UFO crash** occurred the night of July 1 or early morning of July 2 and was discovered by the surveyor Barney Barnett, probably about 10:30 or 11:00 a.m. on July 2. A witness said, "The ground near the crashed UFO was covered with thin pieces of foil."

**1953.** The first **scanning electron microscope (SEM)** was ushered in. With the help of the Engineering Department at Cambridge University and electrostatic lenses, two men, Dennis McMullen and Ken Smith, produced the first micrographs showing the 3-dimensional imaging characteristics of the modern-day **SEM.**

**1989. Inductively coupled plasma mass spectroscopy (ICP-MS).** Samples of decomposed to neutral elements are introduced to a high temperature, argon plasma. This produces four main processes:
  1) sample induction
  2) aerosol generation
  3) ionization by argon plasma source
  4) mass discrimination and detection system

It was the use of the relatively new ICP-MS analysis equipment that allowed the detection of multiple elements on our simple foil shards from the Plains of San Augustin arroyo crash site. This has added a relatively new dimension to the metal analysis. This multi-element analysis can yield results at the extraordinarily minute parts per trillion level. All of Colbern's findings of the multiple element coatings are mathematically reduced to parts per million (ppm).

The timeline of Einstein's 1939 letter to Pres. Roosevelt, to the beginning of isotope research, the development of the scanning electron microscope, through the 1989 ICP-MS invention spans just fifty years. It goes without saying that the equipment and strategies for its use are in the early stages of development. A thorough analysis on the arroyo metal foil shards probably could not have been done as easily before the ICP-MS devices became available.

We do not know the extent of the metal shard foil analysis, as it was found at the site in the late 1940s or 1950s, when the scanning electron microscope was invented. The government had hauled off the complete flying saucer. This no doubt included all of the remaining foil, which was under tough, outer layers. There were, however, at least 15 foil shard pieces left in the soil, which we found.

This prompts a few questions about the analysis of the arroyo metal shard foil: What research, if any, was done in 1947, 1948, or shortly thereafter to match today's sophisticated analysis that we show in Colbern's 2010 report? If the shards were analyzed prior to 1989, what available procedures were used to give them similar results? (The ICP-MS was first used in 1989.) A witness said there was foil on the ground at the crash site. Most of it was apparently picked up, but some was left for this researcher, his friend, some others, and a crew of Navajos. Are there possibly some boxes on a forgotten shelf in a government warehouse with more of the arroyo metal shards waiting for that future day when they might get a more proper analysis? Perhaps. If this is true, then these independent private researchers may be the first to discover the sophisticated and varied coatings on the strange New Mexico arroyo foil.

Considering that our government may be beginning to do analysis with equipment just recently developed, it is understandable that they would be nervous about the security implications of this research and that the scientists who work with and around laboratories with government funding are also reluctant to come forward. Perhaps this research might stimulate government interest in older UFO material it may have in its possession.

# CHAPTER 4
## REFERENCES

1. Campbell, Art; personal correspondence with aluminum companies and metallurgists.
2. Colbern, Steve. *The Artifact Analysis by Dr. Russell Vernon Clark, 2010.* p. 5.
3. Colbern, Steve. *Analysis Report on Metal Samples from the 1947 UFO Crash on the Plains of San Augustine, New Mexico,* 2010.

### OTHER REFERENCES USED

*Aluminum, How it is Made and Where it is Used, Part I. The Story of Aluminum;* The Aluminum Association; 1968.

*The Thomas Register;* Thomas Publishing Co., 510 Plaza, New York, NY 10001; 1997.

Breton, Bernie, *The Early History and Development of the Scanning Electron Microscope,* www.2.eng.cam.ac.uk/~bcb/history.htm, November 14, 2009.

Worley, Jenna and Steve Kvech. *ICP-MS,* www.cee.vt.edu/ewr/environmental/teach/smprimer/icpms/icpms.htm, April 2000.

Two indispensable references were used concerning worldwide aluminum patents. *Designations and Chemical Composition Limits for Aluminum Alloys in the Form of Castings and Ingot* And the companion booklet, *International Alloy Designation of Wrought Aluminum and Wrought Aluminum Alloys,"* 1998.

# CHAPTER 5

## RESEARCHERS, GERALD & THE MOTHERLOAD

Three old navy men in the desert seems like an unlikely scenario, but we were linked together in one of history's ironies. We had brought forth information on one of mankind's greater and more interesting events, an alien space vehicle crash on the Plains of San Augustin in 1947. Many have been reported, wreckage found and analyzed, but none we know of have the complete had an interest in building and enrolled in New Mexico State University, completing a civil engineering degree in 1968. For some years, Chuck worked for a firm that was awarded contracts to build air traffic control towers. Settling back into New Mexico at Gallup, Chuck ran and owned a successful contracting business for over 20 years. After leaving the Navy in 1968, Chuck joined the Naval Seabee Reserves and for the next 26 years worked his way up through the ranks and retired as a Chief Warrant Officer in 1999.

My own background is totally different. I was born in Bonne Terre, Missouri in June 1932. My father was a civil engineer, trained in mining engineering and geology at what is

*Figure 5.1. **Gerald Anderson** points the way to the UFO crash path where important debris was found. In this 1990 photo, Anderson revisits the San Augustin crash site after 43 years. Left to right, Stanton T. Friedman, retired nuclear physicist and long-time UFO researcher; Gerald Anderson, who first visited the site when he was five with family members in 1947; John Carpenter, a psychiatric social worker involved in an early investigation; and Robert Bigelow, financial backer. The crash path was over a quarter-mile long. This is where our research group found over thirty pieces of UFO debris. "Skewed isotopes" indicate the metal shards and heavier pieces found here are not from this planet. Photo credit: Cosimo Books and Don Berliner, co-author* Crash at Corona.

provenance of eyewitness accounts, definite extraterrestrial material found, and laboratory analyses to reinforce the off-planet findings (possibly with one or two exceptions). Who were these men? How did they come to be associated with each other? And what were the circumstances?

Chuck Wade is a local New Mexico man who had grown up in Corona, New Mexico, and went to high school there, graduating in 1957. He joined the Navy in 1959 and served with the U.S. Navy Seabees until his discharge in 1963. Chuck always now the University of Missouri at Rolla. We lived in Willow Springs, Springfield, and later St. Joseph, where I graduated from Central High School in 1950. I had previously joined the Naval Reserves in my junior year of high school. I served in naval aviation at shore stations and later on the *U.S.S. Boxer* during the Korean War. After initial training in food service, I eventually settled in the Kansas City area and prepared for a teaching career. This led me to public school education as a teacher, coach, career counselor, and eventually a high school principal. I retired in 1989 and became interested in the Plains crash site in 1994.

Gerald F. Anderson was born in the Hoosier State at Indianapolis in 1941. Gerald's father was a skilled tool and die maker who worked before and after WWII at Allison Aviation. After the war he worked in the booming appliance business and opened a successful washing machine repair shop that he and Gerald's mother managed. His older brother, located in the Albuquerque, New Mexico area, thought that Gerald's father's skills as a tool and die maker might be useful in the growing scientific and technical fields in New Mexico. Gerald and his family moved from Indianapolis to Albuquerque in the early summer of 1947, where Mr. Anderson found a high security government job at the Sandia Corporation near Albuquerque in 1948.

The family moved around in the area for the next few years. Gerald attended several grade schools, junior high, and high school, and in 1958 he joined the Navy and became a parachute rigger. For obvious reasons, this was a very important position. Today, riggers are responsible for air crew survival equipment, live preservers, maintenance of flight clothing, and survival radios. Gerald, like many in his day, attended college on the G.I. Bill after the service. Most of his education is concentrated in the sciences. He holds an associate of science and bachelor of science degrees in microbiology with an emphasis in molecular biology. He has worked in environmental service in several states. Gerald is now retired and living in northern Green County in southwest Missouri.[1]

After the discord between Gerald and some early UFO writers and others, interest in the crash site greatly diminished. This was two years or so before my first visit, but I did not learn of the friction over the site until I started reading old UFO material, magazines, and books in the late 90's.

My association with Chuck Wade, after meeting him at the 2004 Aztec Conference, was very rewarding, and Chuck had the wisdom to see early on that what we were finding was unique. After we had the Navajo crew to the site in 2004 and 2005, Chuck became very enthusiastic. Living in the Southwest at Gallup, Chuck was able to attend many UFO conferences and extend the knowledge of our finds to many others. Eventually, he got the attention of Roger Leir, of alien implant fame, and his A & S scientific organization. Dr. Leir showed little interest in my original finds, but as more foil started showing up and Chuck made it available, the Open Minds organization decided to finance the ICP-MS analysis. These stunning results can be seen in several chapters of this book. Chuck became the early promoter and salesman for the crash site and its increasing discoveries.

In retrospect, all of us had something to offer. Mine was time, an interest in history, patience, finding the initial material, writing, and site preservation. Chuck brought a burst of energy, which at my age and geographic location in the Pacific northwest was impossible. Bringing in the Navajo crew was a turning point. Chuck took the metal shards they found and let the UFO world know about our site and findings. Bringing Gerald back in with his metal detector was no doubt the most important decision that was made about the crash site.

He had been out of the UFO scene for nearly 20 years. What had transpired to cause this site and Gerald to be pushed back in the pages of history? It began with a simple phone bill on February 4th, 1990. By pre-arrangement Gerald called Kevin Randle, a lead ufologist, and an interview was conducted. A little later a dispute broke out between Anderson and Randle over the length of the phone call. Randle said it was some 52 minutes long and Gerald said it was only 26 minutes. To prove his point, Gerald cleverly altered his phone bill to embarrass Randle. Unbeknown to Gerald, Randle had recorded the call. Gerald sent the altered phone bill to one or two people in the UFO community. They checked with the phone company and found that he had altered it. Gerald immediately was perceived in the UFO community as someone who would go to dishonest lengths to prove a point. The UFO community tolerates some errors including untruths now and then, but expects its UFO witnesses to be impeccable.

Over the next three years, Gerald submitted several documents that did not measure up in investigators' eyes. Some material was sent to forensic specialists who confirmed that one or two of the documents was not what it purported to be. Gerald's supporters dropped away. By 1993-94 he became angry, bitter, and disillusioned and left the UFO scene. This is where the story stood for almost two decades. It turns out the Randle-Schmitt team had problems of its own. Don Schmitt was carrying some very dishonest baggage that would become quite embarrassing. Kevin Randle and Don Schmitt had published two books together – one in 1991 and another in 1994. Schmitt gave an interview to a Milwaukee magazine associated with a large newspaper. He produced a 1990 bio listing his educational background; it was impressive. He said he had a Masters degree from Concordia college and was pursuing a Doctorate in criminology. He also told others, including Randle, that he was a medical illustrator. In reality, he was a letter carrier at the Hartford, Minnesota post office.

When the story broke in 1995, Kevin Randle refused to believe it. He joined into the fray to defend his partner and co-author. He was widely quoted in a letter to ufologists on the internet in February 1995 and on the Rich Planet website. He said, "Of course this rumor [about the credentials] is no more true than the Gerald Anderson story." We are taking another look at this observation in our book.

The reader may ask, why was all the time, energy, and resources spent in meetings, discourses, and numerous letters, trying to disprove what Anderson was telling investigators?' Another question might be, why was Gerald not taken to the crash site with an investigative team, some metal detecting equipment and a shovel or two to locate crash debris? They would have probably found some of what we found in over ten years in working in the site. Our guess is that neither side wanted to gamble on the embarrassment of being wrong. 1) Many were sure the military had picked up everything if debris had been there, and digging would have accomplished little. 2) It was easier to pontificate and present positions in UFO newsletters – and later on home computers – when they became available. Both the UFO community and Gerald became polarized on this issue. Rivalries between competing UFO teams gained momentum very quickly. Key historical individuals that made Gerald's story interesting, including Barnett and the as-yet-to-be-discovered archaeologist, were moved out of the story. (Gerald claimed that one of the Albuquerque High School teachers name Burskirk was the archaeologist.) In essence, Gerald and his Plains story incorporated from his childhood and family memories, plus his additions, became of little interest to researchers. Not because it wasn't true, but there was so much controversy, it was easier to move one than sort it all out.

Whether the reader has followed UFO material for some years or just a few, some of this information is probably on your shelf. You may recall Stanton Friedman's and Don Berliner's very good book, *Crash at Corona* (Paragon House 1992). The Berlitz and Moore 1980 blockbuster book, *The Roswell Incident* (Grosset and Dunlap) should also be available. *Crash at Corona* features a very telling photo (Figure 5.1). It was taken on a gray day in 1990 about 25 miles south of Datil, New Mexico. Four men are looking west and one is pointing. He was Gerald Anderson. The others included Stanton Friedman, Robert Bigelow, and John Carpenter. Anderson had brought them there.[2] He said it was the site of the 1947 UFO crash on the Plains of San Augustin that he and members of his family found. Gerald had learned of the recent interest in the site through a rerun of the *Unsolved Mysteries* TV program in 1990. The previous day a small helicopter was used to find the site. Was this the Plains of San Augustin flying saucer crash that Friedman learned about in Bemidiji, Minnesota, twelve years earlier? It proved to be.

### FINDING THE MOTHER LODE

The majority of our finds so far had been within 200 feet of the small rise where the saucer was said to have come to rest. This ground, for the most part, was easy digging, soft sandy loam, and loose soil, which facilitated screening. Here the gap was well-defined as it cut its way through the buck brush years ago. We had always assumed that the small things we were finding, such as the foil shards, the shoe sole, the wax pieces and perhaps a body part, the split in the craft that Barnett reported. Although we had walked the upper arroyo many times, there were no signs of additional scarring of the terrain from the crash.

We brought in a new, fresh crew after we had introduced Chuck Wade to the landowner and some degree of trust had been established. The author was traveling early in the summer of 2011, and Chuck took the new crew in to do more research. Chuck had become good friends with Gerald Anderson, who had said he was at the crash site in 1947 when he was five years old. Chuck invited him on the June 2011 trip to the site. A key component of their efforts involved the use of Gerald's good metal detector, which immediately began to pay off. More and much larger pieces were found in the few days they were there.

We had always thought the military had scoured the area surrounding the gap and had picked up all the obvious loose crash material, but apparently this was not the case. They may have only concentrated on the trench leading up to the UFO and the bodies, and perhaps were not thorough in cleaning up other debris in the

*Figure 5.2.* ***The first large piece*** *(W-1101) was found using a metal detector on the second day of the 2011 dig. Piece is 15 inches (46 cm) long by 6 inches (15 cm) wide. ICP-MS diagnosed 36 different elements in the metal: aluminum, copper, manganese, magnesium, iron, and silicon were major components. This was a major find and probably an exterior piece of the crashed craft. (Left to right) Gerald Anderson, Nancy Wade, Cara Fay, and Chuck Wade. Gerald was invited back to the site in 2011 and 2012 when the big material was found.*

grazing area and sagebrush. There were signs of a raised dirt berm on one edge of the arroyo close to where the large pieces were found. We did some research on these berms in the area and they were not uncommon. They appeared where there was conservation efforts were carried out, including seeding, land and watershed restoration, and other projects. Scraping this soil into berms inadvertently helped expose buried debris found in 2011 and 2012.

The large pieces, as pictured in the next two chapters, were buried under several inches of undisturbed hard soil. Most of the new, larger pieces that we found were under bunch grass; some were under the soil around the sagebrush. In due time there was a natural buildup of range vegetation more or less obscuring small crash pieces, so they blended into the landscape. Sharp-eyed searchers who knew what to look for found about 15 small UFO honeycomb pieces with or without metal detectors mixed in with the range grass and ground cover. One spectacular piece was found partially under soil wedged under some rocks. After a few years, when the vegetation returned to its natural state, whatever monitoring had gone on tapered off.

When the first honeycomb piece was found in 2011, it was caked and filled with bark dust and debris from a mouse den. (See Figure 5.3.) Also indicating the length of time the piece was buried in the den was over one-and-a-half inches of slow growing moss on one end. This piece, we believe, was possibly

*Figure 5.3.* **Honeycomb artifact in mouse den** *(from above). Bark dust pathway built by generations of deer mice led to artifact. It may have originally been stashed by military cleanup crew. When snow covers the ground, the bark dust trail becomes the bottom of a rodent tunnel. Although covered with debris and feces for many years, some moisture did get in, contributing to moss and lichen growth on the honeycomb.*

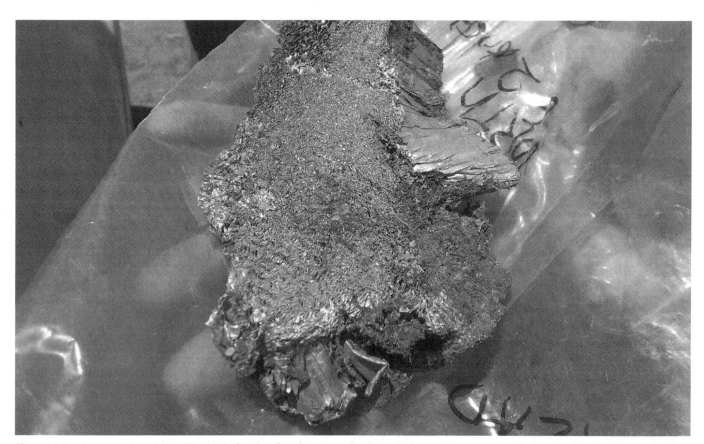

*Figure 5.4.* **Honeycomb artifact** *(W-1102) shortly after discovery. Evidence indicates it was sandwiched between a heavier and lighter piece of the craft exterior. It is 6 inches long (14 cm) by 4 inches wide (10 cm) and it is 1.5 inches thick (3.5 cm). It weighs 4 oz. (126 g). The honeycomb may be part of an energy dispersal system. Channels are compacted with many years of mouse nest debris. Darker area to the right is* Dicranum flagellare *moss. Botanists tell us that for moss and lichen to grow on this metal, calcium and magnesium would need to be present. Both were found in lab analysis. This moss does not grow in New Mexico. Could it have been brought in by the UFO?*

# UFO CRASH AT SAN AUGUSTIN

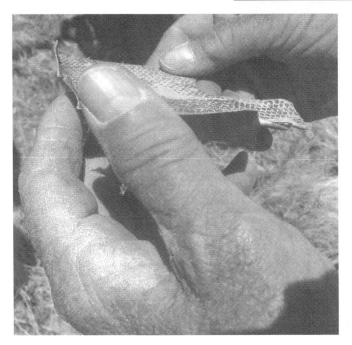

*Figure 5.5. **Another piece** of the light metal (W-1103) was believed to be part of the interior honeycomb skin. It was discovered later that the dots line up with the honeycomb channels.*

## NATURE'S SMALLEST

June has always been a nice time to be on the Plains. The winter show is gone, the winds die down, but there is usually a nip in the early morning air until the sun comes up. What was going to be green that year had started to peek out from shady places looking for the sun and a little warmth. One could see signs of nature's smallest, the deer mice (*peromyscus maniculatus*) around rocks and crevices. They had built up little, brown bark dust paths that were here and there around the rocks. Generations of the small creatures were well aware of their status in the food chain, and had a labyrinth of tunnels under the snow to stay out of view of aerial predators. Now that the snow was gone, only the little bark trails along the rocks were visible on the surface. The mice had gone into deeper tunnels and emerged only at night to forage for food.

Deer mice are often confused with another rodent somewhat larger, the pack rat. Pack rats on the Plains are a hardy bunch, often inhabiting old prairie dog complexes nearby. They

*Figure 5.6. **Honeycomb piece** (W-1102) after it was cleaned. Hexagonal holes go through the piece like a wasp's nest. Honeycomb may have been scraped off of bottom of UFO as it pancaked and skipped once, coming down through a rocky area of the upper arroyo before coming to rest. Artifact is primarily of aluminum; other key metals were chromium, iron, potassium and manganese.*

stashed by someone earlier, as it was nearly one-fourth of a mile from the honeycomb material found later, as well as material we think that came from the bottom of the craft. It was obvious that all of the material described in Chapters 5 and 6 was very old and had been buried or out of sight for quite a long time.

In 2004, Benjamin Shirley of our Navajo crew picked up a cast aluminum piece of material. It turned out to be very important and was heavy enough to be part of the craft exterior. It later became known as W-6. (See Figure 4.11). It was lying on the surface about 200 feet above what we called "the gap." We did not readily have the funds to analyze the metal pieces, and it was several years after finding it and the metal foil that we were able, through the Open Minds organization, to get the initial metal tested in 2010. The test results were astounding; with only some of the material tested, the isotope analysis showed the foil had coatings of many elements and was beyond our 1947 technology. It was one thing to tell those interested that we had found foil shards not known to the aluminum industry but quite another thing to find that they were made with a high degree of sophistication, obviously someplace else. To some, the scientific analyses we had done earlier were just words on paper, but once our material got to the right scientists, it was a different story. Our science reports began to reflect that what we were finding was very unusual, and the basic research showed nothing similar from worldwide aluminum product patents. What keeps many fueled in the UFO enigma is not the endless articles, lectures, or books about this theory or that, but hard, factual, undeniable reports and physical alien evidence that substantiates alien crafts landing or crashing on Earth. What we were developing at the Plains of San Augustin site was hard scientific proof that an alien craft had crashed in the arroyo.

usually live in more permanent residences marked by an exterior pile of wood chips they collect and put in and around their dens. It is believed that on one of the author's first digs in an old prairie dog complex nearby, pack rats (known for bringing odd attractive things to their nests) brought in a variety of things, including a foil shard with isotopes not matching any on Earth. (See Figure 3.4).

Literature on deer mice indicates the female has up to four or five litters per year. If this nest were continually used since 1947, over 240 generations of the little creatures had nested in and around the honeycomb. Figure 5.4 shows the honeycomb just after extraction from the nest with bark dust still clinging to the exterior. The Plains deer mice are about 3 to 4 inches long with a tail of at least equal length. Deer mice have large eyes adapted to night vision. For warmth in cold weather during the winter, they often sleep in large groups to absorb and preserve body warmth. They survive well on plant greens and are partial to seeds. They will also eat larval insects, lichen (which is plentiful), and occasionally meat when available. They stockpile some of the foods for winter. The snow white underside, large ears, and tan or gray pelt, with large black eyes led one researcher to quip, "It is quite a handsome mouse, as mice go." We agree.

## THE HONEYCOMB

The June 2011 trip to the site, led by Chuck Wade, included seven people. The morning they arrived at the site, about 8 a.m., was very eventful. After completing some camp chores, a couple of members of the group took a metal detector over to a rocky slope not far from the gap. I had found a little shoe sole there in 1997, some years before. This morning, they were working that rocky area atop the previously described mouse winter tunnels. They noticed quite a bit of the fine bark and juniper needles in a rock crevice, which had obviously been an under-the-snow winter nest for rodents. They ran the detector over the rocks and got a low whine. They moved the detector to near the entrance and the whine got louder. They moved it to the right, and the whine subsided; they moved the detector to the left, and the whine subsided again. They moved it back to the entrance, and the whine again intensified.

## NO BEER CAN

It is not unusual to find metal pieces in and around rocks and old prairie dog mounds. In the Plains arroyos, the local mouse population, of course, has nothing to do with this, but the larger, previously mentioned pack rat certainly does. An acquaintance of mine, who is retired from the Bureau of Land Management, had a small showcase in his office of items he had scrounged from pack rat nests. These were flat round rocks, part of a logo from an old Frigidaire ice box, various kinds of foil (fingernail size), and a sardine can lid neatly curved around the opening key, but the centerpiece of the entire collection was a $20 gold piece (circa 1913) from a ramshackle one-room sheep herder's cabin. A photo from the cabin showed the nest overflowing the oven compartment of an old stove.

Our group's hit with the metal detector in the rock crevice was interesting but not unusual. After WWII meat shortages, there was a general livestock boom in the area. Good prices could be had raising cattle and sheep. Many of the local ranch hands frequented a tavern on Highway 12, not far away. The beer cans apparently have mirrored these good times with high sales in canned beverages. Trash in rural America, for many years, was an "out of sight, out of mind" affair. Trash in gulches and ditches near any habitable buildings in the West bear this out. What could be burned was put in a burn barrel (nowadays mostly illegal) or in a heating stove to help start the morning fire. Cans with any residue inside attracted rodents and other undesirable vermin. What to do? What they had always done, in New Mexico: dump the cans in the arroyo down the road, and after the next rain or flash flood, they magically disappeared.

Of course, we were all aware that some aluminum beer cans were in the arroyo, even in rocky areas. Sometimes they were found after someone set them up after target practice. The two searchers knew enough not to reach in for the shiny metal, out of respect for their hands, snakes, and other unseen dangers. They had brought along a trowel and started poking in the old mouse den. Later one of the crew members quipped, "It was too thick to be more foil, and too square to be an aluminum can, and the edge grooves looked unfamiliar."

The quart or more of brown bark dust material that they scraped out of the rock cleft was apparently a very old deer mouse's winter nest. By the time they got most of the debris off, they could see the shiny edge and some of the honeycomb structure. This was no beer can! They shouted, "Hey, we've found something!" They certainly had.

We know the site had been monitored for some years. Whether the piece had been found and hidden in the crevice by a site monitor in those years, or it had been stashed there by a member of the Army clean-up team (in hopes of retrieving it someday) is open to speculation. I personally think it was stashed there by a military person for later retrieval. It is unlikely it made its way into the crevice by itself when the UFO crashed.

Quite a bit of the ground-up bark from the rodent nest and some moss was still adhering to the honeycomb artifact as it was slipped into a plastic bag and labeled. Several larger pieces of flat metal were found some distance away in the next several days. We discovered later that these pieces had a direct relationship to the honeycomb and the craft propulsion.

As mentioned before, the first piece of heavy flat metal was found in 2004. It was found above the gap, where we found most of the smaller material — some wax, the HDPE artifact, the shoe sole, and the important pieces of foil discovered on previous trips. Also found were fabric scraps and the wafer. Detailed analysis of the foil was described in the previous chapter. These other items are described in Chapter 7.

*Figure 5.7. **Large pieces of metal**, very hard and very strong, found at a depth of three inches. Insert above shows the piece being uncovered. This material may have been driven into the ground when the craft made its first impact and skipped down. We were curious about the brown dots on one side of the metal. Later they would become very significant.*

## SKIPPING IN

We had always assumed that where the craft ended up on the rise and had cut a 100-foot furrow into the arroyo brush and sandy loam soil was where it had spilled most of its findable contents. Barnett mentioned the craft had a split in it when it came to rest. We are not sure if the split was vertical or horizontal. We surmise that these rather delicate shingle-like pieces of foil were in a protected place (above the UFO centerline). The split Barnett mentioned may very well have been vertical if these delicate pieces of foil were forced out of the aperture prior to the craft coming to a stop. How several small beings made their way out of the craft and onto the ground after the craft came to rest is a mystery.

In *The Plains of San Augustin Controversy July 1947* that Stanton Friedman had sent me around 1999, there was a transcription of a telephone interview in February of 1990 between Gerald Anderson and Kevin Randle, a leading UFO investigator. For the sake of brevity and clarity, some of these statements were taken out of their original context. Anderson's answers and comments to Randle revealed two things we already knew and shed considerable light on things we were yet to learn. On the final slide before the craft came to rest, Gerald said, "It looked like it had bounced a couple of times, cut a furrow and buried itself."[4] We had found the place Anderson had alluded to when he talked of a "furrow." We called it the gap, and there was little growth of what should be thick buck brush in it. An extensive soil analysis indicated that the soil on either side of the 30-foot-wide, 135-foot-long "gap" was different but similar to the soil in the middle. A Catron County extension agent came out to the site and suggested it was indeed fill from nearby.

Anderson said other things that resonated about the site. Randle asked him to describe the debris. Gerald said, "Like pieces of metal. Like somebody had thrown around pieces of tinsel. If you can imagine an explosion in an aluminum factory...that's pretty much what you had in that area."[5] Jesse Marcel Jr., whose father brought home some of the Roswell debris, recalls trying to piece them together on the kitchen floor. He said it was "mainly some kind of thick metal foil." Was this a reference to what Debbie Ziegelmeyer, the Missouri MUFON director and Frank Kimbler, the Roswell teacher, found at the Roswell site? It quite possibly is. We had been finding shiny thin foil metal on and under the sandy loam for years. Some of it still appears on the surface especially after a rain. Chuck and his Navajo crew had found some in excavations in 2004 and 2005. The Navajo foil was analyzed in 2010 and found to have multi-coated isotope combinations that could not have been manufactured by human engineering.

The final item that would not be substantiated until 2011 and 2012 was damage to the bottom of the craft's hull. Continuing to describe the debris in the Randle interview, Gerald said, "I

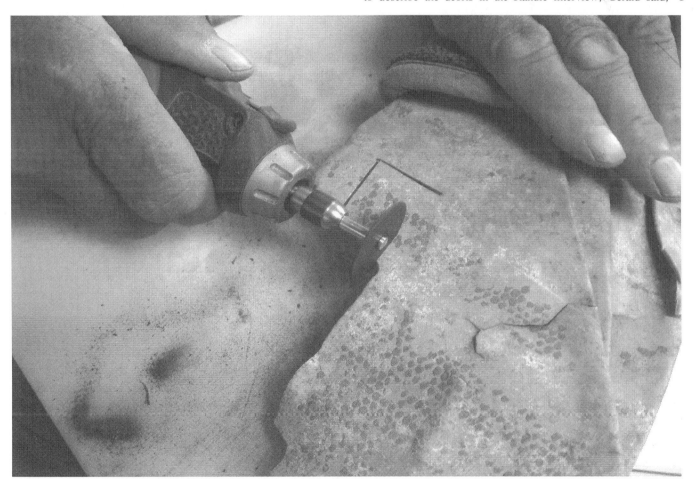

*Figure 5.8. **Chuck cutting the heavier metal** (right) with the brown dots to send off to the lab. Samples from all finds at the site, including the metal, are dispersed to many labs and individuals around the world, including South Africa and Hong Kong, for safekeeping.*

don't know what the part looked like that had impacted and was partially burned, but apparently the underside of this had sustained a lot of damage."[6] We had no evidence or debris to indicate the craft bouncing and causing damage to the underside. We thought it was a logical observation and let it go at that. All of that changed when Gerald came to the site with Chuck's 2011 group. Much of the material he found in the upper arroyo had obviously come off the bottom of the craft. We had always thought any skipping or bouncing must have happened in the lower arroyo. Gerald took several people to another site we had not worked and began finding material immediately. Those pieces are featured in this and the next chapter. Gerald was the only witness we knew of who mentioned that the craft "had considerable damage on the bottom." Gerald was there...he had to have been there or had intimate knowledge of the site to lead others to it and have all the details right.

Do I think Gerald's recollections were colored and added to by his older brother, cousin, uncle, and father? Certainly they were. Who among us has not had family recollections altered, shaped, or blended concerning past events? This is especially true if same-generation siblings, cousins, or other relatives were present and there is a discussion of shared events. This family had unique circumstances to keep the event just among themselves. We will learn later that Gerald's dad worked for a company doing high-level government research. Good, secure jobs like this were hard to find. This did not stop the family from talking about the arroyo crash, although it was discouraged.

**MILITARY MINDS**
One would assume that the military would be all over the flight path the saucer took on its way in to crash. They may have been searching for surface debris, but they apparently left debris buried in the soft sandy loam of the arroyo that had buried itself as the craft came in. Some agricultural leveling equipment used by the military to help fill in the trench no doubt covered other UFO debris, but what we found was very shallow.

Conservation practices as recommended by the Salado SWCD had considerably slowed erosion higher up in the mountains, and it is believed man-made items were beginning to "erode out" of the sandy arroyo, as less sand washed in. In 1947 the nearby highway was gravel. One or two old-timers who lived in the area recalled a road grader that was stored a few miles from the arroyo that belonged to Catron County. It was rumored that the Army borrowed it sometime after the war to clean up some kind of missile crash. A search of the Catron County archives turned up nothing.

There was also the problem of metal civilians' trash items being buried in the arroyo sand, which would have caused considerable difficulty for the military if metal detectors were used to find buried crash wreckage. In the late 1940s the military might probably have considered the "out of sight, out of mind" theory. It is doubtful they could comprehend the UFO movement beginning 25 to 30 years later and that there would be an interest in any crash here because it was probably considered "a godforsaken place." Those of us who have worked this site sometimes still share this view, albeit with a little reverence. History tells us that the primary objective of our secrecy those days was to keep technology from the Russians. Several Soviet spies were already active in New Mexico's Los Alamos laboratory in 1947 and would help facilitate the very early detonation of Russia's first atomic bomb 25 months after this UFO crash, in August of 1949.

The readers may recall that the author flew over what was assumed to be the crash path for several miles before and several miles beyond it. No unusual patterns were visible from the air. One wonders why the military did such a slipshod job of lightly covering the gap trench and skip-down place with a thin layer of soil, which quickly eroded away. The emphasis may have been on getting the bodies and craft out quickly (in the first day or so) and covering up any trench possibly caused when the craft skidded through the soft sandy loam soil where the foil shards were found.

We should think back, and if the Plains crash were discovered on July 2, 1947, it was probably thought of as a one-time event. We know that Mjr. Jesse Marcel did not get to the Brazel-Foster Ranch debris field until July 6th or so. Contact with the Brazel family in 2013 indicates that some think Mack found the UFO debris about July 4th. The military crew at the Plains crash may have been inexperienced in these matters, and there is a good possibility they had not been trained thoroughly. The need for this training would probably not be apparent until after the Roswell crash debris was picked up by Major Marcel a few days later.

If special units existed for alien crash retrieval, they apparently were in their early stages of formation and training. We should bear in mind that all branches of the military between 1945 and 1950 (when the Korean War began) were in a state of transition from WWII to a peacetime footing, give or take the Cold Ware considerations. In retrospect, in the middle of this period of flux, there were several well-known UFO crashes, including the Roswell scenario, and the lesser known Plains of San Augustin crash about the same time. Another crash that many believe was alluded to in Frank Scully's 1950 book, *Behind the Flying Saucers,* was an incident that occurred in 1948 near Aztec, New Mexico.

In December 1947, the U.S. Army Air Corp broke away from the Army to become the U.S. Air Force. Another factor that may have contributed to inexperience of the military personnel is the need for constant training and upgrading for any units who may have been assigned to alien crashes, along with a high turnover in lower level enlisted personnel in those days. A good example of this inexperience would be the threats and intimidation that Barnett and other witnesses experienced at the Plains crash site. These were people clearly not trained to deal with civilians.

**THE HONEYCOMB SANDWICH**
In the June 2011 trip to the crash site, as the crew worked its way up the arroyo, finding occasional large pieces and some smaller ones, we began to find at least one side of each piece had some rather unusual brown dots on the surface. The other surface was clean. At the end of the first day or so, it did not take us long to notice the honeycomb's hollow spaces matched up with the dots on the recently found large pieces. Each large piece (See Figure 5.11) had the brown dots only on one side. We surmised that the honeycomb material emitted or carried some kind of power, and the pieces of metal with the dots, which we later called the skin, covered the honeycomb.

We think some large pieces of the craft bottom were

torn loose when the craft hit high ground, possibly several times some miles north of the arroyo. We are sure the craft skipped down at least twice in the arroyo before coming to a stop. If the craft had pancaked elsewhere and lost part of the lower half of its external shell, including some of its power structure, it may have dropped pieces between the first and last impact in the arroyo on the way in, some of which the team found. The first honeycomb, if it had been stashed by someone, was close to where the craft eventually came to rest. The larger flat pieces were found up the arroyo in what we assumed had been a flight path of the craft.

Incidentally, the small, approximately 10-12 foot rise the craft came to rest against was the last obstacle of any height for miles. If it had cleared this, although damaged, the momentum may have taken it another 400 yards or so. Apparently, what we had were pieces of the outer three layers of the bottom of the craft. The honeycomb was sandwiched between the heavier and lighter pieces of metal and we believe, backed by an I-beam grid.

The honeycomb surface was quite fragile, and many of the tiny aluminum channels (entrances or exits), which we believe went entirely through the honeycomb structure were bent over or misshapen. An occasional perfect, six-sided, even-walled structure remained intact. We assumed this was the ideal functioning shape, as we could also see some strange brown dots to match on the inside of both top and bottom coverings. Since the honeycomb channels were slightly larger in diameter than the brown dots, we assumed these upper and lower coverings were flush against the honeycomb channels, similar to a tight-fitting skin. One of our researchers originally surmised that the brown dots were some kind of adhesive holding the honeycomb on the skin. Another theory was that the brown dots were residue from some interaction in the channel. We eventually found the dots to be completely independent and connected to a power source of the craft. See Figure 6.6.

## THE SHIKOKU ISLAND HONEYCOMB

Through Internet research we were able to locate only one mention and photograph (See Figure 5.9) of another honeycomb, allegedly from a crashed UFO in 1971. The event occurred on the evening of February 26 on the smallest of the Japanese islands, Shikoku. This island is known for its ancient history, temperate climate, beautiful scenery, and is primarily an agricultural area, known for its fruit. The population of the island is just under two million people. There was no significant WWII industrial production on Shikoku. Some of its picturesque scenery and 81 temples were featured in the 1980 TV mini-series *Shogun*. Shikoku Island has four main divisions (states or regions). Our story takes place in Ehime prefecture on the west side of the island, facing the Seto inland sea about 500 km west of Osaka and about 200 km south of Hiroshima.

According to the book, *Alien Honeycomb*, "It seems a Mr. Shiota, who lived not far from the city of Matsuyama, saw a bright orange, curiously shaped ship plunge down behind a mountain visible from his house. The ship burst in air with great noise and fell. The next morning he went to the mountain and looked for the place where the ship had fallen or crashed. He found a great area where the ground was burned and trees scorched and knocked over. On the ground, he could find only one piece from the wreckage." (See Figure 5.9.)

*Figure 5.9.* ***Japanese honeycomb*** *found on Japan's smallest main island, Shikoku, in 1971. Note vertical channels and dot pattern below the honeycomb. Also note the intact covering or skin adhering to the top of the honeycomb. Compare this honeycomb to the one we found below and the honeycomb we found in the next chapter. Above photo from Jun-ichi Yaoi archives, Nippon Television.*

*Figure 5.10.* ***Honeycomb match?*** *These brown dots line up with the large honeycomb piece's hexagon channels. The team also found lighter metal with some dot configurations. (See artist's drawing on the next page). At the time we had idea who they were.*

## Artist's Drawing of Honeycomb "Sandwich" Configuration

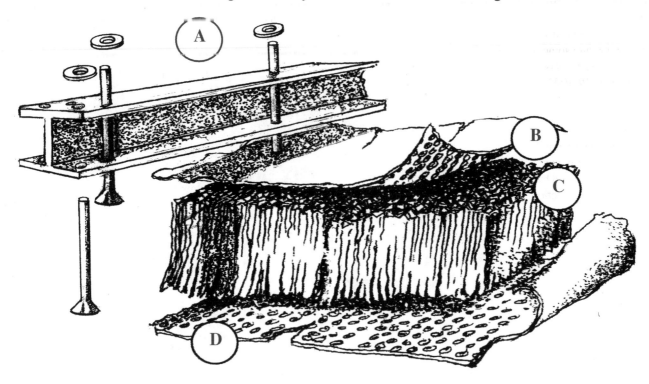

*Figure 5.11.* **Honeycomb "sandwich" with H-beam backing:** *A) H-beam is believed to have supported honeycomb from inside the craft. Flanged holes would have allowed fasteners to be flush against honeycomb and interior W-1103 skin. B) Interior W-1103 interior skin with brown dots facing down. Major components are aluminum, silicon, iron, zinc, and manganese. C) Honeycomb W-1102 is primarily an aluminum alloy layered structure in thin sheets formed in hexagonal channels. D) Believed to be outer skin of craft W-1101 and much heavier than the skin above, the outer skin also has dots facing inward, which are thought to be part of the power generation process. The above beam (A) would curve slightly indicating a basket-like web of curved beams backing the honeycomb skins on bottom of craft. It is believed most of the craft's bottom contained this honeycomb and brown dot system.*

The man who covered this story at the time was Jun-Ichi-Yaoi, a well-known producer of UFO documentaries with Japan's Nippon Television Corporation. John Pinkney, an Australian UFO researcher and author of *Alien Honeycomb*, asked Mr. Shiota whether anyone had analyzed the honeycomb piece. Jun told the author that he didn't think so, saying, "Mr. Shiota does not wish to be made fun of by scientists."[7] The Japanese producer went to a shop where the honeycomb, which is about the size of a card table, was located and photographed it.

**EVIDENCE OF REPAIRS**
We kept finding evidence in the wreckage of the Plains' crashed UFO that occasional repairs had been made. Our first indication was a small hole in our first heaviest piece of flat metal (W-1101). Tests on the isotopes told us it was not an earthly alloy. We later established it was torn from the bottom of the UFO. There were basically two sides to the metal. One was a relatively smooth side we assumed was on the exterior of the craft. The other side had those strange brown dots (see figures 5.9 and 5.11.)

The small hole in W-1101 was flanged or countersunk on the opposite side of the dots. We think that some kind of flat, aerodynamic-head fastener was inserted from the outside of the craft to bring a section of dots closer to the honeycomb (W-1102). For better functioning, the heavy metal (W-6) had no brown dots on it when it was found. We think this may have been an outer skin of the upper exterior of the craft where another or modified power source was used.

We compared our honeycomb and brown dots with the pattern left on the skin or covering of the 1971 Japanese Shikoku Island crashed UFO. We saw no break in the dot pattern under their honeycomb (see Figure 5.10). Ours, of course, had many years of weathering and exposure to the elements. Some of our honeycomb skin had been buried for 60 years or so, but the dot pattern remained perfectly intact. When we found pieces of the skin on the surface where the dot pattern was worn away, we would occasionally find a drilled countersunk hole on the outside of the skin. This indicated to us that some kind of a flush-headed fastener (such as an aircraft rivet) was going into the craft from the outside. Later we found the I-beam with the same-sized holes going into it (see Chapter 6). We finally considered the inevitable: our Plains craft had seen better days and those that sent it out were taking a calculated risk to crew and craft that it may not come back. Apparently some kind of aerial mishap had damaged the craft and it had been unable to return to its place of origin.

When we found the I-beam, it was too crude to have been a new UFO part and was obviously a repair. The holes that were drilled in the ends were hardly precision and looked hastily done. It would seem out of place that the I-beam was part of an original, new, highly functioning craft. More than likely, it was a repair piece for an older craft. We think that a grid of beams like this may have been a backing to support the honeycomb, although we are sure some more sophisticated system of joining the gridwork I-beams was used. We also have reason to believe the lighter metal W-1003 with its brown dots facing down was possibly backed by a square, open gridwork of small beams this size, we think 10-13 inches square. It would seem logical that any lattice work built into the honeycomb had originally been some kind of metal fusion. In the repair, the repairers apparently could not use a fusion process or high heat, and the I-beam had to be secured and spliced by hand with some kind of drilled hole, flange, and pin construction. This is not to say the work was shoddy, but custom to the job at hand and as strong as it could be using a different but effective process for the repair. We believe the builders had access to the entire aluminum I-beam structure similar to a basket with beams horizontal and vertical. However, the aluminum honeycomb was no doubt added to this structure in later stages of this construction. The honeycomb probably reached across much of the underside of the craft. Structural damage on the I-beam frame would have been very hard to repair with the exterior honeycomb in place. We will go into some detail in the following chapter about the role of the brown dots on the inside of the skin coverings and some stunning research concerning them.

Many of us see this field in possibly a more favorable light than we should. Not all UFOs are perfect machines, and they may not function well in some earthly conditions. Severe weather, being shot at later with surface-to-air missiles, malfunctioning guidance systems, and pilot or crew error may have led to many crashes on Earth. Finding a craft that is less than perfect may not be as uncommon as one would think. Indications are that our arroyo craft was less functional than it had been is almost reassuring. We reuse space vehicles many times and there have been some disasters. Remember the *Challenger* disaster of 1986 and the *Columbia* re-entry breakup of 2007. Who alive in the seventies can forget the drama of Apollo 13 as the crew coped with an oxygen tank explosion that caused an unlikely chain of events? Because the Roswell and Plains crashes may have been on the same evening, is it possible the aliens took a gamble and sent a less functional craft to assist the Roswell craft and both were lost? We wonder.

### OTHER CONCERNS

We have written a great deal about the metal foil we found, but little has been said about the wax pieces. It is believed from a very few people who have been allowed inside the upper compartment of a recovered saucer (Bob Lazar said he was allowed to look in one) that there is an insulation coating sprayed on all metal surfaces. It would make sense if this material consisted of some of the wax pieces we found and had analyzed. The military did what they could to get the downed craft out of the arroyo, filled in a trench over 100 feet long and inadvertently covered some of the vital components with sandy loam soil from nearby.

One complication at the site occurred apparently three

## SUMMARY OF STEVE COLBERN'S METAL SAMPLE ANALYSES

**THE METAL SAMPLES: SCIENTISTS' CONCLUSIONS**
**Metal artifacts from 2011 Dig.** Extensive analysis was done on these 2011 finds. Our analysis included testing by Inductively Coupled Plasma Mass Spectrometry (ICP-MS). This test also included analysis for both trace element content as well as isotopic ratios for the three samples. Up to 30 different major and minor elements were found in each sample. Only about a dozen were significant, major elements, listed in the graphic below. Some isotopic ratios were determined using minor trace elements not listed.

**MAJOR METALS**
Aluminum
Manganese
Copper
Zinc

**SAMPLE W-1101: HEAVY METAL**
Our analysis by ICP-MS from the San Augustin site indicated the isotopic ratios for selected elements (boron, magnesium, nickel, copper, and zinc) does not correspond with known aluminum alloys available from 1947. **The isotopic ratios of nickel and zinc did not correlate with known terrestrial values. This sophisticated metallurgy was not available in 1947.**[8]

**MAJOR METALS**
Aluminum
Chromium
Iron
Magnesium

**SAMPLE W-1102: FIRST HONEYCOMB**
The elemental analysis from the crash site indicated that the composition of the material did not correspond to that of any known aluminum alloy that was available in 1947. The isotopic analysis also indicated that **the isotopic ratios of the nickel and copper in the sample were sufficiently different from the terrestrial values and was probably not made on Earth.**[9]

**MAJOR METALS**
Aluminum
Copper
Manganese
Iron

**SAMPLE W-1103: LIGHTER FOLDED METAL**
We have done extensive analysis of the metal known as W-1103 believed to be the interior skin of the San Augustin crashed UFO. We ran the sample through the ICP-MS process and tested for trace element content. We were able to find and isolate five selected elements (boron, magnesium, nickel, copper, and zinc). The sample did not correspond to any known 1947 alloy. **The isotopic ratios of the nickel and zinc were different from terrestrial values. This leads us to believe the material did not have its origins on earth.** One isotope of magnesium in this sample (M26) was also present in an amount **significantly greater than in terrestrial materials**.[10]

weeks after the arroyo crash. About the time the site should have been cleaned up, trench filled in, craft and bodies and all signs either buried or taken away, a flash flood came down the arroyo. The water at the Highway 12 culvert was about four feet deep, and it washed out the of the approaches to the culvert. The highway was temporarily closed. Our guess is that the flash flood uncovered some of the debris and played general havoc with the newly filled-in trench. This might have made the military realize how vulnerable the crash site was to severe weather and precipitation. The possible monitoring of the site was discussed in an earlier chapter.

As we will learn later, Barnett was queried and harassed by military representatives to keep quiet and not divulge his UFO crash findings to anyone. I think at first Barney was very selective about who he talked to about the crash discovery. Why then did it seem he was singled out for warnings and visits to his home about keeping things quiet? Barney may have had more information about the disposition of the craft; for instance, he may have let the military know about the rail head loading area that was still open in Magdalena. Incidentally, a connection of this old ATNSF north-south railroad crossed park street one-half block from Barnett's house. Once he retired in 1959 and was no longer concerned about keeping his job, those in charge worried that he might have been talking about the incident more.

We surmise from Chapter 3 data that the site was monitored until at least the mid-1960s. Our team in 2012 found some pieces on the surface in the upper arroyo. It is an irony that a dedicated civil servant and WWI veteran may have contributed to lowering of the lower arroyo sands through encouraging good conservation and reduction of topsoil runoff by nearby land owners. By just doing his job, the artifacts that came to the surface (some of which we found), gave generations that succeeded him a glimpse of the cosmos and an answer to one of life's great questions: Are we alone? Thanks to Barney Barnett and his fellow conservation workers, we have our answer.

# CHAPTER 5
## REFERENCES

1. Campbell, Art. Personal correspondence with Gerald Anderson 1996–2012.
2. Friedman, Stanton T. and Don Berliner. *Crash at Corona*, Paragon House, New York, 1992. p. 70.
3. Op. cit. Correspondence with Gerald Anderson.
4. *Plains of San Augustin Controversy, July 1947: Gerald Anderson, Barney Barnett, and the Archaeologists*. J. Allen Hynek Center for UFO Studies, Chicago, Illinois, Fund for UFO Research, Washington, D.C., p. 63. 1992
5. Ibid.
6. Ibid.
7. Pinkney, John and Leonard Ryzman, *Alien Honeycomb: The First Solid Evidence of UFOs*. Pan Books, 1980. p. 117.
8. Campbell, Art. Personal correspondence with chemist/materials analyst Steve Colbern.
9. Ibid.
10. Ibid.

### RECOMMENDED READING

Carpenter, John S. "Gerald Anderson: Truth vs. Fiction." *MUFON UFO Journal*, no. 281, September 1992

*Plains of San Augustin Controversy, July 1947: Gerald Anderson, Barney Barnett, and the Archaeologists*. J. Allen Hynek Center for UFO Studies, Chicago, Illinois, Fund for UFO Research, Washington, D.C., 1992

# CHAPTER 6

## THE I-BEAM, SECOND HONEYCOMB & PROPULSION

**65 YEARS LATER**
The rules were simple, "If you find anything on the surface or underground, mark it with a red flag and call for assistance." The first important find was about 10:30 in the morning on the first day of the 2012 expedition. John LeMaster found it wedged under an overhanging rock and one below it. When the cry went out, everyone knew it was unusual. The party was scattered doing their

on that bright spring morning on the Plains of San Augustin.
"Hey, I found something!" a man's voice called out excitedly, "Come over here!" John LeMaster had been working a new area in the upper arroyo with Gerald Anderson and one or two others close by. Everyone stopped what they were doing and funneled towards where the others were running. Several who came up the arroyo were out of breath when they arrived. John pointed, "There, under the rock." Peeking out of the shade of an overhanging rock and into the sunlight was the end of the I-beam with a red flag posted nearby. Heard from the gathering group was, "No, it can't be. My gawd! Is that one of those...?" Another voice said, "I-beams?" Another said, "I don't believe it." Symbolically, this was one of the most important finds at the site. Jesse Marcel Jr. had mentioned

*Figure 6.1. **I-beam found on 2012 trip to Plains site** was much larger than any researcher had imagined. Length was 13 $^1/4$ inches (33.6 cm); bottom and top flanges were 2 inches (5 cm) across. Total height was 1 $^1/4$ inches (3 cm); weight was 13 ounces (368 grams). This is the first photo taken. Note undisturbed dirt with juniper needles and twig on the lower part of the beam. Drilled and countersunk holes indicate this had been a repair. It was not memory metal. Arrow inside box indicates twisted and burnt end coming out from under dirt.*

search patterns over a 200-yard swath. When they converged on the find, everyone knew it was significant. Renee Carkin recalled to the author, "It was bent up, twisted and burned on one end, but it was absolutely beautiful." An ex-Roswell Army-Air Force intelligence officer had described some similar but much smaller I-beam material he found from the Roswell debris field in 1947. Jesse Marcel was being interviewed by Stanton Friedman in 1978, some 34 years ago. To our knowledge, no one had ever found another I-beam at a UFO crash site. All that changed 65 years later

one in 1947 that his father had retrieved at the Roswell debris field. More surprises were to come.

Those gathered there began jockeying for a good angle for digital pictures. Someone, in order to get a better photo, picked it up and was immediately reminded about the rules: "Flag it and leave it." It was quickly put back in the rock crevice. Nancy Wade and Murff arrived after it had been picked up. Their job was to document all finds. Murff was visibly upset and Nancy was none too happy. When the I-beam was removed from the crevice,

# UFO CRASH AT SAN AUGUSTIN

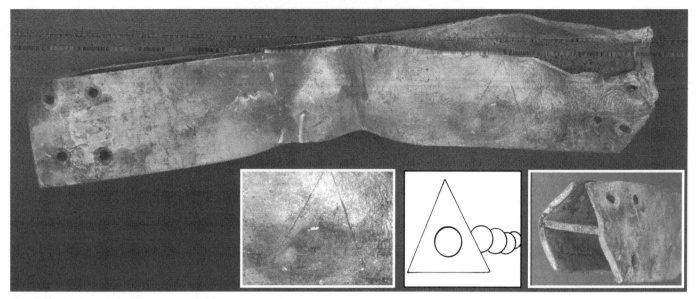

*Figure 6.2. **Structural H-beam**. It is well weathered and oxidized. It is believed to have come off UFO when it skipped in the upper arroyo. Burnt end, upper right in the photo suggests high heat or explosion. Metal composition is primarily aluminum with high concentrations of silicon, manganese, copper and iron. Minor traces of 60 other metals were found. Metal composition matched nothing found on earth. H-beam may have been in a grid structure of 12-13 inches squares behind and supporting the honeycomb on the bottom inside of the craft. Inserts: Left detail of circle/triangle; center, diagram of sky path or eclipse; right, left end of beam.*

some caked dirt was wiped off. In their zeal over the finds, some enthusiast had upset the chain of documentation very necessary for a UFO artifact on the surface. We were able to locate one digital photo of the beam before it had been picked up. Clearly, it had been there for some time. Packed, dark dirt with juniper needles covered a lower section of the beam, where the triangle/circle logo was found later. See Figures 6.1 and 6.2.

After about 20 minutes or so, Renee Carkin was asked to pick up the beam again so others could see it from different angles and photograph it. She related, "It was lighter than what I expected...it was about thirteen inches long with a dent in the middle, which surprised me after having read about the light but sturdy unbendable nature of the Roswell materials that Marcel had discovered." We are sure some of the Roswell craft was a later technology and quite possibly a different propulsion system altogether. Renee continued, "It was generally dull in appearance. It had a slight glisten to it, and when I held it up to the sunlight it looked slightly iridescent." Another man wrote, "Even the skeptics now believed something had crashed there." He also said, "Every time we found something, it seemed to energize all the others, and I remember looking down into a shady area at empty lawn chairs where some of the women had been sitting."

It was inevitable that skepticism would surface in such a large, diverse group. UFO researchers see it frequently, and for the most part, it is healthy. Skepticism and denial that something exists when you are holding it in your hands is something else. One of the key Roswell writers looked at the remains of what we think was an artificial body organ (discussed in the next chapter) and said it looked like a collapsed weather balloon. It is one thing to cherish fiction and spread it as truth, but a definite line is crossed when you delude yourself into believing it. A few years ago, he was pictured in *UFO Magazine* standing in the arroyo crash area and was quoted as saying, "This is a bogus site." Over his left ear was where I had found the item I call the HDPE artifact, and where the Navajo crew and others found metal shards that according to skewed isotopic tests were foreign to this planet. To his right, a hundred yards or so and some five years later is where John LeMaster found the I-beam. Bogus site, indeed! If he had brought along a metal detector and a shovel, he could have probably found something important.

## I & H-BEAMS

The I-beam has been around for centuries and was used by ancients who worked primarily in stone. It is believed to have developed on earth from a single piece of stone or wood placed in a doorway with two side posts as supports. The I-beam of today is usually steel and was developed at the beginning of the Industrial Revolution in the 1800s. Pre-stressed steel and concrete I-beams can be seen in modern highway construction. I-beams of laminated wood are often used to span large rooms like auditoriums and gymnasiums.

Our modern steel I-beams have their history in railroad rails and framing throughout iron ship construction. The beam we found at the crash site is actually more in the shape of an H. The I-beams, which are used to hold up vertical objects like skyscrapers, are known as UCs (universal columns). UC I-beams have equal or near-equal width and depth. Our beam is in more of an H-shape, which is called a UB (universal beam) and has deeper and longer flanges to carry a horizontal load that may occasionally bend or flex. The H-beam we found at the crash site we believe was in a bowl-shaped grid system which may have supported the entire honeycomb structure from inside. The H-beam was strong enough to support tremendous G-force and stresses as the craft maneuvered, sped up, stopped, or turned at speeds of from 6,000 mph to 8,000 mph.

## THE TRIANGLE & CIRCLE SYMBOL

This symbol faintly appears near one end of the I-beam. Frankly, it appears to be nothing exotic unlike the hieroglyphic-like writing reported to be on a much smaller I-beam found near Roswell. Jesse Marcel, Jr., who saw the wreckage his father brought home in 1947, recalled the symbols were embossed on the I-beam in a purple-pink-colored substance. This symbol message no doubt had meaning to the builders of the craft that had obviously dropped the I-beam. Figure 5.11 shows it in a possible configuration in our craft. UFO fans and readers may recall that Jesse Marcel, Jr., was one of the first civilians to pick up a UFO I-beam. His father brought this home from the Roswell debris field.

The discovery of the triangle-circle logo was exciting. The later find of the receding orbs may give this logo some cultural significance also. We think the logo was stamped on the beam at the time of manufacture, possibly when the metal was still warm. The design may depict the type of facility or repair, or even the place or race of alien origin.

A close inspection of the drilled and flanged holes tells us that it does not look like anything remotely approaching precision work. Nor do we think the craft was made with rivets, screws, or bolted together. More than likely this piece was a repair and for some reason could not be welded or fused like the original construction. We believe it came from the bottom of the craft and may have been part of a basket-type or shallow bowl or grid that supported the honeycomb structure from behind. At one time, the flanged holes may have been guides for flat-headed fasteners that were flush with the surface of the beam. It probably lay flat on something else, which would obscure the triangle/circle symbol. Following this line of thought then, it may have been a backing for one of the two flat metal surfaces with brown dots that we found in the wreckage. It was obvious that the metal with the brown dots faced each other. We have designated the W-1101 and W-1103 as an exterior and interior metal surface with the honeycomb sandwiched between them. We think the I or H-beam was mounted on the inside as a framework to support the honeycomb. The triangle/circle logo may have gone against the I-beam flat surface and this surface was mounted against W-1103, the light, metal skin of the honeycomb as seen in Figure 5.12.

## SKY PATH & ECLIPSE SURPRISE

During the photo layout for these pages, we had cause to enlarge the triangle/circle photo. We had darkened the triangle lines with graphite so they would show better. The graphite brought out four or five partial circles coming from behind the lower right side of the triangle as seen in Figure 6.2. They were overlapping and graduated in size, very similar to what is known as transits or orbital sky paths. These are seen in huge crop circle formations found in fields all over the world; the primary and most elaborate are seen in England. The first records of crop circles appeared in Europe in wood cuts about 1647.

We decided to research the orbs lined up behind the triangle. We saw lots of beautiful and elaborate, full aerial views of what apparently were celestial bodies going through various phases. Our view of the partial circles receding behind the triangle seemed unique. It was almost as if our overlapping orbs were from an artist's perspective view of a planet (if that's what they were), coming towards the triangle from a distance.

I happened onto a book by Linda Moulton Howe, written in the millennial year of 2000. She had some news of apparent crop circle planetary bodies overlapping taken by various photographers in the 1990s, but they referred to these overlapping orbs as eclipses. I thought, "of course" this made sense. What I apparently had coming or receding from behind the triangle were faint lines in various stages of an eclipse culminating with the full circle in the triangle. See drawing in Figure 6.2.

Thinking to get a better photo of these orb lines, we had a professional photographer take a close-up of the triangle-circle and no engraved lines of the eclipse orbs were visible. Nevertheless, they were clearly visible looking at the photo taken in natural light. We assume here that the I-beam had been in a high-heat environment. These perfectly curved parallel lines may have been some form of paint or stain, but they clearly were not engraved.

Whatever goals and agendas the aliens have for us, if any, is occasionally overshadowed by their appreciation for beauty and precision as evidenced by the many genuine crop circles found all over the world in the last generation. We, like many researchers, see evidences of more than one group of aliens and they may be competing with each other with their design and manifestation of their considerable skills in this important phenomena. What more inspiring and beautiful event can compete with a total eclipse? This is one of the few events all of us share in the universe.

Some of us may not have been present at an eclipse but the media has made them available to all who care to view the awe-inspiring event. It was probably no coincidence that UFOs were filmed at eclipses visible from Mexico in 1991, 2008, and 2009. It is comforting to this writer that the appreciation of creations and the science behind them are shared by other beings living, working, and traveling in the universe if this is an eclipse symbol. It is also exciting to share with the reader (if the crop circles were authentic) what we believe may be the first markings obtained from UFO debris and their connection to similar designs and themes in worldwide crop circles.

*Figure 6.3. Logo from Rendelsham Forest UFO found in 1980. Sketch is copied from an official USAF report drawn by Staff Sgt. Jim Penniston.*

## THE RENDELSHAM FOREST CONNECTION

The year 1980 was a benchmark of information about UFOs coming to planet earth. That year, we learned about the famous Roswell Incident and read about Barney Barnett's Plains of San Augustin story from the published *Roswell Incident* of that year. A few months later, a UFO apparently under remote control landed in Great Britain near the twin NATO bases, Woodridge and

Bentwaters. Security guards saw a strong light in the forest and three USAF airmen went to investigate it. They reported a craft that was about 3 meters wide at the base in a pyramid/cone shape and about 2 meters high. It had a blinking red light at the apex with blue lights below. Staff Sgt. Jim Penniston made a drawing similar to the one above for his report when he was describing the markings he saw on the object. The elements of a triangle and circle (orbs) are similar to our I-beam logo (Figure 6.2) although they are different configurations. Both logos, we think, might be representations of celestial events. The I-beam partial circles behind the triangle may be of an eclipse. The Rendelsham logo may represent an orbiting body around another larger body, perhaps a sun. Is there a connection between these logos on two UFOs coming to earth 33 years apart — one in 1947 and the other in 1980? We wonder.

## SECOND DIG CREW

Who were these people clustered in the arroyo looking at the strange metal object? One of the team e-mailed, "There were people from across the country with a variety of education and professions. They came from five separate states, and one from Canada. They worked, for the most part, as a team with the common goal of uncovering crash debris. Some had the attitude of 'Oh, sure. Crash debris.' When they left, all I talked to were believers." The dig Chuck and Nancy organized was a very well thought out plan, and it included many people Chuck was in contact with and had met at various UFO conferences. There were Randall and Celene Santeler from Las Vegas and Seth and "Annie" Feinstein who live between Denver and Colorado Springs. There were Tom and Nancy Banks from Las Cruces, New Mexico area; and another New Mexico man, John LeMaster from Santa Fe. Most of the participants were retired, but the youngest was a Navajo, Murff Tilden, who was attending graduate classes from the UNM facility in Gallup. Murff had also been on the 2004, 2005, and 2012 digs, and helped find most of the metal foil shards featured in Chapter 4. Rounding out the group were two women, Renee Carkin from Deming; and her friend Cynthia Kropp, living near Truth or Consequences, New Mexico, due east of the Plains. The Canadian was Mark Olsen from Winnipeg, Manitoba.

Of course, the organizers of the group and veterans of several digs of the Plains crash site were Chuck and Nancy Wade from Gallup; and co-leader, Gerald Anderson from Southwest Missouri. One woman said, "They were terrific organizers." Some grumbled that the trip had been "over-organized." But all could see later that the organization was efficient and considered the safety and personal needs of all participants. The group numbered 15 in all, staying at Socorro's Econo Lodge forty miles east of the Plains site.

## SITE HISTORY, EARLY FINDS

How had this group come together? Chuck Wade had been showing and talking about what we had been finding at the site at some UFO conferences. These were people who had expressed a serious interest in the work we had been doing. I had started finding artifacts myself as early as 1996. I invited Chuck to the site in 2004 after meeting him at an Aztec UFO Conference. We sealed the friendship and later formed a partnership with an A&W root beer. It was not an auspicious beginning — just two old Navy men meeting in the high desert with hopes and dreams that the world would be a better place if we found evidence of alien visitors. A simple goal, but extremely complicated to accomplish, to say the least.

How had the new section of the arroyo that became so productive been found? It goes back to the year 2011. Chuck had gone to the site with six or seven people, including Gerald Anderson. Gerald had been at the site as a child, and 43 years later had taken several men, one or two vehicles, and a helicopter to the site. The helicopter was financed by a Las Vegas real estate executive who was interested in UFOs and the paranormal. Anderson was invited on the 2011 dig at the site. Also included was a young woman named Cara Fay. Cara had also brought along a girlfriend. Included for a half-day was Frank Kimbler, a teacher from Roswell. I had some traveling to do that summer, and at the age of 79 had begun to slow down a bit. When I got home, there was a message on my computer: It was from Chuck. It began, "Art, you won't believe what we found."

In the book, *Crash at Corona* (Friedman and Berliner, 1992), is a photo of four men Gerald had brought to this site in 1990. In the photo (see Figure 5.1.), Gerald is pointing to the exact location of the UFO crash path and area where we found the spectacular debris. We had been finding debris and metal foil shards since about 1997. We did not have the resources or opportunity to get the isotope tests run until 2010, as described in Chapter 4. We knew after Colbern's first report in 2010 that the metal shards had not originated on earth. Chuck invited Gerald to the site in 2011 and again in 2012. It was with Gerald's knowledge of the site and metal detectors that we were able to start finding the larger debris found in these pages.

How had the military left so much material to be found later? We do not think the military did its homework or consulted the right people. They probably did not realize that the average annual rainfall on the Plains in those years was over 13 inches (much less now). They did a fair job of covering up the site, filling in the trenches and skip-down area. As mentioned before, we believe they may have had civilians do some of the leveling and fill-in work at the site. It seems that the goal was to smooth everything over and get it covered, not realizing that a thin layer of dirt would blow or wash away in time. We found evidence of monitors having been there for some years, no doubt picking up crash debris that had worked its way to the surface. Chuck Wade and I found foil shards neatly stacked up, folded, and apparently put back in the sand. A good example of this can be seen in Figure 4.5 in Chapter 4.

What of the great, highly touted, romanticized view of military units crawling on their hands and knees in precision lines, picking up every tiny scrap of crash debris? We saw it on TV re-enactments; we read in many books how efficient these specially trained units were. Was this an accurate portrayal? Was it true? Yes, but probably much later than 1947. Our arroyo was sandy loam, loose dirt, with some large rocks on the side. There were thick, spiny buck brush, prickly pear cactus, and old fragile prairie dog complexes now abandoned. (When stepped on, the tunnels collapsed). There were lizards, pesky fire ants, and occasional rattlesnakes hiding out from predators, including eagles in the old prairie dog tunnels. This was their environment, and we were the alien visitors. No doubt the military learned

*Figure 6.4. Second large chunk of honeycomb as found in 2012. Material as it may have been configured in the bottom, outside the UFO. The bottom would have been covered with an exterior skin of heavy metal (W-1101). This piece has honeycomb skin metal still attached (W-1103) with traces of brown dots on the underside, above right. This artifact may have been damaged and buried by range-leveling equipment in the 1980s. Note the perfect, six-sided honeycomb channel ends on the bottom of the piece.*

from this and subsequent crash sites. But this unfriendly terrain would have been very, very difficult for an inexperienced military unit the first time out. Then we add the maxim of Murphy's Law, "If anything can go wrong, it will." We do not know the unit, probably based somewhere in New Mexico, but it must have been difficult coping with what we believe may have been the first post-war UFO crash, complete with alien bodies that witnesses described.

## LOSS OF POWER

The terrain at the crash site was deceptive. I had always thought the craft that crashed there had been aiming for the miles of old, relatively flat lake bed just behind a 10 or 12-foot-high rocky ridge. If the craft had been flying a dozen feet higher, it would have made it. We believe whatever was controlling the craft – whether manual or some auto pilot arrangement – must have sought flat ground on which to land. As the craft came closer to the Earth, we believe it skipped down one or two times and started dropping vital components of its double-walled honeycomb propulsion system, including the I-beam in the upper arroyo, where Gerald guided the two parties in 2011 and 2012. Whether the power system was at all functional when the craft was less than 200 feet in the air, we do not know. Large pieces of the bottom section of the craft, including several honeycomb pieces, were found in and on hard-packed earth 200 yards from where it finally came to rest. In the lower arroyo weather, erosion and the ever-creeping sand and soil of the alluvial fan eventually showed us signs and artifacts of the alien vehicle crash.

In the short run, the military was able to alter the crash site cosmetically to something similar to what it had been before. They did the best they could. Important things were taken care of. The crashed craft was taken out in a day or two. The trench it had made through the buck brush was filled in. The bodies were removed, but apparently some parts were buried that could not be transported (see Chapter 16). We think the military unit, with the help of some civilians, made attempts to recontour the landscape with various types of local equipment. The ever-changing site was monitored for a time, including the upper arroyo crash area. Eventually, the scars healed and the range grasses and scrub brush grew back. The leveling equipment apparently disturbed the buck brush roots in the arroyo, which left us with what was known in the lower arroyo as the "gap." No doubt, the military people involved finished their terms of service, honored their pledges of secrecy, returned to civilian life, found jobs, had or continued families, and the arroyo kept its secret for over fifty years. One contingency that could not have been planned for was the development of and eventual civilian market for metal detectors.

## SECOND HONEYCOMB

Seth Feinstein was working the northwest side of the arroyo about 11:30 in the morning on the second day. Suddenly, he shouted, "I've got a reading over here!" Chuck, Nancy, Gerald, and John LeMaster immediately went over to where Seth was. There was a strong signal on the metal detector. It apparently was a large object, but not near the surface. John took a shovel and started digging straight down. The ground was hard. Every few inches he dug down, Seth would pass the metal detector over the hole again...It still registered but no artifact was visible.

A large rock lay on the surface, about 90 percent of it buried. The metal detector was passed over it, and the beeper

## SAN AUGUSTIN HONEYCOMB & SIGNIFICANCE — BY STEVE COLBERN

The performance characteristics of reported UFO/flying saucer type craft clearly indicate that they must utilize a type of propulsion which involves the generation and control of gravitational fields. This is evident from the numerous witness reports of these crafts doing right-angle turns when traveling at multi-Mach speeds, accelerating out of sight in a fraction of a second, and generally doing the type of high-g maneuvers which would crush the occupants in a more conventional type of spacecraft.

The details of the process require some clarification, but many accounts indicate that very strong magnetic fields are involved in the production of the inverse gravitational fields that allow these craft to perform as they do; this is the reason for the failure of electrical equipment often reported in the vicinity of a UFO. Some reports, leaked from classified government projects, have also indicated that powerful magnetic fields can influence the inertia of objects, as well as their susceptibility to gravitational fields.

Microwave energy has also been reported being emitted by many of these craft (3 GHz-1,000 GHz); this may be utilized for the amplification of the effect, as certain materials have been reported to move when subjected to powerful microwave beams, via an induced Biefield-Brown gravitational effect, which was first observed in high-voltage capacitors. Interestingly, aluminum is one of these materials.

The effect is also said to be amplified by the presence of magnetic materials, such as rare-earth elements. This may account for the presence of the unusual amounts of rare-earth elements detected in the aluminum materials found at the San Augustin crash site. This effect of microwaves producing a thrust on certain materials was reportedly utilized in a classified government project called Project Skyvault, in the mid to late 1950s.

In the summer of 2011, a piece of an aluminum honeycomb material was found on the plains of San Augustin, New Mexico; the site of the crash of a UFO, which crashed at about the same time as another craft that crashed north of Roswell, New Mexico in early July, 1947 (Figure 5.4). More of the honeycomb material was found in 2012. Many other pieces of debris from the San Augustin crash had been found previously, several of which have been analyzed by this author and others, but the honeycomb pieces are highly significant finds, for several reasons. The material is composed of an aluminum alloy, which was similar to other pieces from the crash site analyzed previously, but had a layered structure, which was made up of thin sheets of the alloy arranged in a lightweight array. The structure also contained hexagonal channels, which ran through the thickness of the piece.

Each side of the piece was covered with thicker sheets of the same material (called skins,) both of which were covered on the side facing the honeycomb with a regular array of "brown dots" (Figure 6.5), which were composed of an organic polymer, containing relatively large amounts of silicon, aluminum, iron, nickel, and the rare-earth elements dysprosium and terbium. Dysprosium and terbium are interesting because they can be used together in extremely efficient mechanical transducers, which turn magnetic energy into mechanical energy; the earliest patent date in human science concerning transducers in 1976 was by the U.S. Naval Research Laboratory, well after the 1947 date of this crash.

The brown dot material was intimately connected to the aluminum skins, as if to make intimate electrical, or mechanical contact and may well have been the material that produced the microwave energy for the propulsion of the craft, utilizing the magnetic energy present to excite the aluminum skin of the craft at microwave frequencies, producing gravitational thrust. The regular arrays of fibers, or wires, inside each brown dot make this possibility more likely. This type of microtechnology was far beyond the state of the art in 1947, in any event.

The channels in the honeycomb piece appeared to be arranged such that there was one brown dot transducer array at each end of each channel. It is possible that, in flight, a fluid, possibly a conductive gas, was circulated through the honeycomb channels to maximize the effect and cool the transducers.

Much more analysis will have to be done on the honeycomb piece and the skin containing the brown dot arrays, but this new data indicates that these pieces probably came from a craft that utilized unconventional, most likely inverse gravitational propulsion, which was far beyond the technology that any nation in the world possessed in 1947.

*Figure 6.5. **Honeycomb with embedded pine needle** (right center) indicates that some or much of the craft's bottom skin (W-1101) was missing as it passed over and apparently clipped a pine tree. The needle must have been lodged in the exposed honeycomb at that time. Note the moss in the middle of the needle, which obviously began growing at the time of the crash. BLM conifer experts believe the needle came from a pod of a southwestern white pine or a ponderosa pine. Maximum elevation of these trees is 7,000 feet. The nearest pines of this type are approximately 35 miles north of the crash site.*

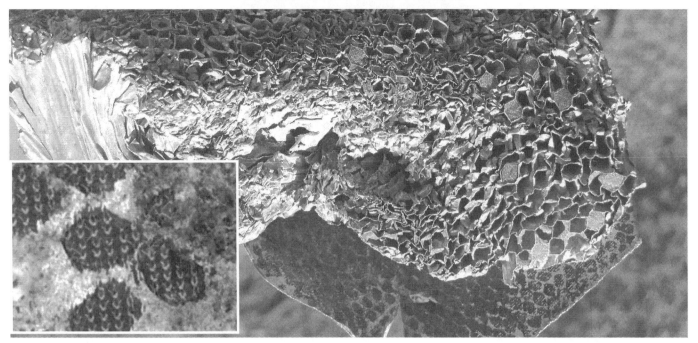

*Figure 6.6.* **Open honeycomb structure is** *at the top of the photo. Immediately below is the honeycomb skin with the brown dot pattern facing up. Magnification (inset) of brown dots at 18 X. Another skin was on top of the honeycomb and had dots facing down. Many thousands of these brown dot honeycomb channel interactions are believed to have produced the craft's gravitational thrust.*

started registering louder towards one end of the rock. John widened the hole and went a little deeper. A strong signal was still emitting from the hole. They were about ten inches deep. The soil was still very hard-packed, but John persisted hacking away with his shovel. He picked up a blade spade and started digging around the edges of the rock. Finally, it was loose enough to pry up with the spade. A glint of metal in a large clod of dirt came into view. someone said, "I think it's another piece of honeycomb." A male voice said, "We don't want to damage it."

One of the men got down on his hands and knees and scraped around the object with a trowel. Finally a shovel was placed under it, and it came free. Seth ran the metal detector over the empty hole and said, "I get no more reading." By this time, Murff and several others had gotten there. He and Nancy were not going to take any more chances on failing to get this documented. The shiny clod was laid on the ground near the hole. Nancy started photographing it with her camera. Murff Tilden began scribbling notes. Chuck said something to the effect of "Take all the pictures that you want, but don't pick it up." This time no one did.

The big piece was seven- to eight-inches long and four-inches wide in an elongated shape. With the packed dirt inside, it weighed about three-fourths of a pound, and like the other honeycomb, it was one-and-a-half- inches thick. Evidence indicates that the craft lost a significant part of its outer skin some distance away, exposing the honeycomb with a subsequent loss in power. As the craft came into the arroyo, we believe it hit at least twice, leaving the material found in this chapter and several other pieces of honeycomb and metal scattered on the arroyo surface.

When John and Seth got the piece out of the hole, someone said, "Hey, it's got metal attached to it." We had found some of this same metal loose before, but this was the first time we had found it attached to a honeycomb. Nancy told someone making a video that we were sure the earlier metal with the brown dots was a skin that had covered the honeycomb. Even though this dirt clod with the honeycomb inside was buried under twelve inches of hard-packed earth and a rock, it still had a growth of moss on it. The moss was brown but still growing although out of the sunlight. As we examined the clod and began to look more closely at the new honeycomb, we could see where some of the skin had been wrenched and torn off. (See Figure 6.4). We believe this may have happened as some leveling equipment passed over the buried artifact and

## WHAT ARE RARE EARTH METALS?

Rare earth metals are on the Periodic Table of the Elements. There are twelve of them. The average person has seldom heard of them. They include Eu (europium), Yb (ytterbium), Dy (dysprosium), and Tb (terbium) as our scientist explained previously. Rare earth elements are not really rare, but they are seldom found in commercial quantities. Most of the world's commercial rare earth substances today come from China.

What are they used for? Everything. Some colors on your TV contain these elements. They are found in catalytic converters in your car's exhaust system. Almost anything mechanical, or electronic, including your computer, uses them. In 2009, China exported over 250,000 tons around the world. After 2010, dysprosium sold for $212 per pound, up from $6.27 per pound a few years earlier.[2]

# UFO CRASH AT SAN AUGUSTIN

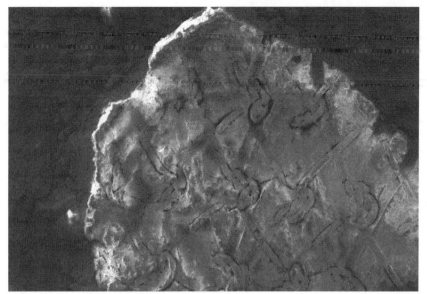

*Figure 6.7. **Scanning electron microscope** (SEM) image of a single brown dot at magnification 48x. Flat appearing, ribbon-like lines criss-crossing the magnified dot are actually microscopic wires (like fishnet) going around the dot. They are literally embedded into the metal skin behind the dots to ensure solid contact. It is thought when the honeycomb skins are energized, the brown dots inside the honeycomb channels interact producing energy.*

the movement of the rock may have caused some of the skin to peel off of the honeycomb, slightly tearing a little of the skin on one edge.

## CRAFT PROPULSION

We have asked our chief scientist Steve Colbern to give us a summary at this stage of the research of some materials found at the crash site and their possible use or function. One of the last items run through the ICP-MS process were the strange brown dots found on the inner surfaces of some metal. They were made with a carbon-based polymer, similar to the HDPE artifact as seen in Chapter 7.

Each brown dot contained 13 elements mixed in a polymer. Six of these were major and consisted of aluminum, manganese, iron, and silicon. Two additional elements were the rare metals dysprosium (DY) and terbium (TB).[1] Dysprosium was discovered in 1886. It is used today in hard disk drives and to make control rods for nuclear reactors. The metal was not isolated in a pure form before 1950.

Terbium was discovered in Sweden in 1843[2]. Carl Mosander was a Swedish chemist who had discovered other elements. While looking at one substance, he discovered two previously unknown substances. One of these he called terbium. This element is a vital ingredient of magneto-restrictive alloys. One primary use is in loud speakers. Terbium is also used in actuators in naval sonar systems and sensors. It is a silvery white rare earth metal, soft and malleable enough to be cut with a knife.[3]

## BROWN DOT PROPULSION THEORY

Scientists suggest energy generation occurred in the honeycomb channels. At each end of the channel, two brown dots on the covering skins faced inward. ICP-MS analysis of the dots gave us all the ingredients but no recipe. It is believed that in flight, a liquid or conductive gas was excited in the honeycomb by magnetic energy, providing microwave frequencies and gravitational thrust. Criss-crossed lines like tiny wire-ribbon grids were wrapped around the dot and embedded in the metal of the skins (W-1101, W-1103) to ensure contact. The two rare earth metals, dysprosium and terbium, worked together and were excellent transducers, converting magnetic energy to mechanical energy.

Paul E. Potter, another UFO propulsion researcher says:

*Most of the structures of these craft are made from aluminum or aluminum alloy, sometimes laminated with other materials to aid their conductivity or storage of electric charge, and sometimes formed with a **honeycomb** structuring for strength and lightness.*[4]

Potter talks of a fluid going through what he calls a "toroid." We assume that the toroid is a series of parallel channels (like our honeycomb) about 1 ½ inches thick. He says:

*Contained inside the toroid is an electrically polarizable fluid that when rotated round the toroid generates a magnetic field that extends far outside the craft — exactly what that fluid is contained inside the toroid is subject to further research, possible it is a gas in the form of*

*Figure 6.8. **Renee Carkin** admiring the I-beam, described earlier, found in 2012. She was one of the most capable and enthusiastic of our Plains team. She is a graduate of Boston's Emerson College in mass communication and broadcasting. She calls herself an energetic grandma from Deming, New Mexico.*

*deuterium gas (deuterium can be electrolyzed out of water as a "heavy hydrogen" gas and separated from the oxygen of water), which can be ionized by passing an electric current through it. It could be that the toroid holds a liquid as simple as water modified in such a way as to be insulating, but carrying metallic suspensoids so as to generate electrical charges.* [5]

He also writes about another potential liquid and says: *Another possibility might be that an electrolytic liquid containing metallic particles could be used, and propelled inside the toroid's insulating walls so as to induce extremely high voltage electrostatic charges, by interface charge separation, laminar charge separation, or triboelectric charging.*

Potter goes on to say:
*In all cases the toroid would need to be made of an insulating material, perhaps a metal-like aluminum laminated with insulating skins. This would be necessary to prevent electric charge leaking away from the inner fluid through the toroid's casing.* [6]

It is clear that Potter is on the right track. His statement above is very close to what we found in our brown dot/honeycomb analysis. Potter said, "Their conductivity or storage of electric charge was sometimes formed in a honeycomb."[7] We are not sure what medium was in the honeycomb channels (if any). It may have been a gas. Potter mentions the possibility of deuterium gas, which is "a heavier and stable isotope of ordinary hydrogen."[8] Potter also suggested that a liquid as simple as water could carry metallic suspensions to generate electrical charges. Apparently, our double-hulled craft (with the honeycomb sandwiched between) generated power.

Steve Colbern, our scientific advisor, thinks some energy may have "excited" the metal of the top and bottom

## SCIENTIST'S CONCLUSIONS: SUMMARY OF ANALYSES OF I-BEAM & BROWN DOTS

**I-BEAM ANALYSIS 2012 SECOND DIG:.** Again ICP-MS and ratio analysis was done on this important find. Tests showed 70 metal traces total in the I-beam. Tests show 37 were in the metal originally, and 33 were added by aliens. Highest level metals over 500 ppm (parts per million) are seen in the list under major metals below.

**BROWN DOT ANALYSIS FROM 2011 & 2012 DIGS**
ICP-MS tests revealed the base material of the brown dot to be a polymer. Mixed in with the polymer were several major metals. The two rare earth metals seen below work together to turn magnetic energy into mechanical energy.

**MAJOR METALS**

Aluminum
Manganese
Silicon
Iron
Copper

**I-BEAM SAMPLE W-1204**

We have done extensive analysis of the I-Beam found at the San Augustin, New Mexico UFO crash site. The analyses performed included testing by Inductively Coupled Plasma Mass Spectrometry (ICP-MS), which included analysis for both trace element content, and isotopic ratios for selected elements (magnesium, nickel, and copper).

The elemental analysis indicated that the bulk of the material was composed of aluminum, with approximately 1% magnesium, 0.3% silicon, 0.2% copper, 0.2% iron, and various other trace elements, including chromium, manganese, gallium, calcium, sodium, potassium, titanium, and zinc. A total of 33 elements were detected.

**The isotopic analysis indicated that the amounts of the heavier isotopes of nickel, and copper in the sample were sufficiently different from the Terrestrial values to lead us to believe that this material may be of extraterrestrial origin.** [9]

**BROWN DOT COMPOSITION**

Polymer with
Aluminum
Silicon
Iron

**BROWN DOT SAMPLE**

Analysis was done on the material from rows of brown dots, which were present on one side of sample W-1102. The analyses done on the sample included Energy Dispersive X-Ray (EDX) bulk elemental analysis, and Inductively Coupled Plasma Mass Spectrometry (ICP-MS) trace element analysis.

The brown dot material contained mostly carbon, melted under the electron beam used in the EDX testing, and appears to be an organic polymer that contains knotted fibers and metallic elements as additives.

**RARE EARTH METALS**

Dysprosium
Terbium

The main additive elements included 0.7% silicon, 0.2% iron, along with relatively high concentrations of rare-earth elements dysprosium 1,380 ppm (0.13%) and terbium 500 ppm. It was discovered that dysprosium and terbium are used together in modern mechanical transducers; this may therefore be part of the function of the brown dot material. Other elements detected included sodium, aluminum, calcium, zinc, and selenium, along with traces of gold and platinum. A total of 40 elements were detected in the brown dot material. [10]

honeycomb skins, causing an antigravity reaction in the millions of honeycomb channels. The brown dots were virtually embedded (scientists call it "intimate") into the skins, assuring good contact. It is not known what source of energy may have been used to "excite" the honeycomb skins.

I asked Steve what he thought energized the brown dots at the end of each honeycomb channel. He said, "It was almost certainly electrical energy." He thought that the dysprosium/terbium transducers in the brown dots were most likely to generate vibrations at certain low audio frequencies, which have been said to cause an anti-gravity effect. Steve Colbern called this process "inverse gravitational propulsion." Potter describes,"metalized parallel plates 10-15 mm apart," much like our honeycomb. He says, "The hull of the craft could be used as a lens for focusing electromagnetic waves." The honeycomb obviously made power. Whether the craft was assisted by additional energy forces is not known.

We are not assuming here that all UFOs have this propulsion system. There may be many systems that were used in the past and today to propel alien craft in our skies and atmosphere. We think it appropriate, however, to note that what you are reading here and much more was known by the U.S. government by early fall of 1947.

## CHAPTER 6
## REFERENCES

1. http://en.wikipedia.org/wiki/Dysprosium
2. http://en.wikipedia.org/wiki/Terbium
3. http://ngm.nationalgeographic.com/2011/06/rare-earth-elements/folger-text
4. http://www.linux-host.org/energy/ufogravity.htm
5. Ibid.
6. Ibid.
7. Ibid.
8. Ibid.
9. Colbern, Steve. *Summary Report, I-Beams*, January 1, 2013
10. Ibid.

## RECOMMENDED READING

*Mysterious Lights and Crop Circles* by Linda Moulton Howe, Paper Chase Press, 2000.

# CHAPTER 7

## OTHER FINDS
## "THAT AIN'T NO PART OF NO COW"

I recall in an earlier trip, late one afternoon, driving into the restaurant at Datil's Eagle Guest Ranch, the Plains were gathering shadows in the low places. The tops of distant mountains to the east were still bathed in sunlight, and all of the mountains in the Plains basin were taking on the soft mauve and purple of evening. As I stepped out of the truck, a cool wind reminded me to reach back inside and get my denim jacket. I tugged my hat down and headed toward the corner door of the restaurant.

The screen door, with an eerie screech, announced my arrival. After I had closed the inner door and my eyes adjusted to the dimly lit interior, I could see six or seven people. A young couple was seated near the window, hunching over their table concentrating on each other and peering at a joint menu. A man was sitting on a sturdy barstool at the cedar-paneled bar. His mannerisms were familiar, especially the way he tipped his head back as he drained his can of beer, showing his weather-worn stockman's hat with a narrow black band.

A huge dusty white longhorn steer head glowered from the age-blackened fireplace at two poorly supervised kids, squabbling over a plate of French fries. Suspended from an aged pine board ceiling were an oxen yoke, a wooden hand-cranked washer wringer, an old McClellan army saddle and several singletrees hanging with assorted western memorabilia. At one end of the room was a case of Native American artifacts, and at the other end a humming, burnished stainless steel pie case. Near the door was the stuffed head of a mouflon ram, (the locals pronounce it mufon), and on a wall nearby, as if looking for a place to run, were the mounted heads of two pronghorn antelope. Another window table was occupied by two older men. One was tall and lanky, the other a heavier, darker man in his mid-1950s. The latter of the two sported a dark mustache and a short stubble of beard. His partner had a rather pleasant square leathery face. Both had that tell-tale, lily-white area on the upper part of their foreheads and an indented band of compressed hair around the back of their heads where their hats had settled in. One could tell these men had spent a significant amount of time in the sun and weather. They both wore pin-striped, white, long-sleeved western shirts and sported gold badges worn by livestock inspectors of the state of New Mexico.

We had exchanged the customary western nod when I sat down, took my hat off, and draped my coat over an empty chair. I pulled the artifact out of my coat pocket and laid it on my blue and white bandanna in front of me. I could tell the men had ordered and were waiting for their food. I thought, well, this investigation has to start somewhere. I picked up the artifact, walked over to their table, and asked them if I could get their opinion on something. "Sure," the tall one said, "what ya got?" I laid the artifact on the table. "Where'd ya find it?" asked the short stocky one. I said, "Down the road a piece in an arroyo leading onto the Plains." He reached for it and picked it up. "Huuuuuh!" he said, as he rubbed his thumb across the bottom. He turned it over looking at the golden brown side. "Looks like she's been in a fire or sumpthun," he said as he handed it to his partner. His partner turned it over in his hands and said, "I never seen anything like it." I said I was wondering if it might be a petrified part of a cow or afterbirth or something? The lanky one said, "That ain't

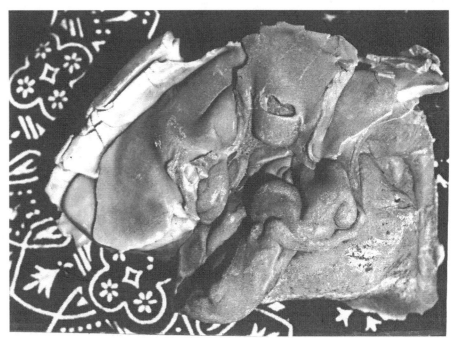

*Figure 7.1. **Top side of artifact on author's bandanna** as it was found half buried in the arroyo sand. It had apparently been soft, and then hardened in the elements. Medical experts thought it might have been a pump of some kind. FT-IR spectroscopy analysis shows the three layers are made of an amorphous high density polyethylene material without a crystalline structure. Found in the arroyo gap in 1995, it touched off a fifteen-year investigation and research.*

no part of no cow!" His friend chimed in, "That ain't no part of anything I ever seen." About that time the waitress set down two huge salads smothered in ranch dressing. As the artifact was handed back to me, one said, "Sorry partner, we can't help ya." And the other chimed in again, "Never seen anything like it." I returned to my table and put it back in my pocket. I thought, if these guys who spent their lives around cattle and ranches do not recognize it, at least that probably rules out the possibility that it was connected with local livestock or wildlife. Somewhere off in the background some western music was playing. It had been a fruitful ten days, and I would be going home tomorrow.

The earlier part of this trip to the Plains had not been that fruitful. Twelve days earlier I had experienced overheating of my engine about the time I got to Seligman, Arizona. I was able to get water every few miles until I got to Holbrook, when I turned south off of I-40. I thought I would try to make it to Datil, New Mexico, where I had a reservation for the night and see if I could go into Socorro the next morning to get a repair. I stopped at stations, now closed, and obtained some water from the service islands and supplemented it with water from my cooler and a five-gallon jug I had with me. After I left Springerville, Arizona, it was very hilly. Every time I went up a grade, the needle on my temperature gauge went up also. I finally limped into Datil about 11 p.m. and was greeted at the guest ranch by several, not-too-friendly barking dogs. I had arranged for the motel people to leave my room open. I found it and settled in. After the dogs quieted down and a coyote stopped yipping on a nearby hill, I had a good night's sleep.

The next morning after breakfast, I decided to stop at a site which I had identified on a previous trip. I had enough water in the truck to cool the engine when it got hot, and then I would go to Socorro that afternoon. When I reached the arroyo basin, I decided to take a shortcut and turned down the old water course. It looked firm, but that was deceptive. I was stuck for several hours, gunning my truck back and forth in the soft sand with the overheated engine, unable to get anywhere. To make a long story short, I had walked back up the arroyo and flagged down a local man who pulled me out with his truck. I spent the next day in Socorro getting a new water pump and getting back to the Plains.

All of this had set my plans askew. I had been gone six days out of a fourteen-day trip and had spent a good deal of money on the water pump and some other unexpected expenses. I had in my truck a rough camping outfit, and by filling the cooler with deli sandwiches, ice, and other assorted quick foods, I was pretty self-sufficient. The man that had pulled me out of the arroyo was remodeling an old house on his property, and I was invited to stay there for a night or two. This put me much closer to the area on the Plains where I was working. I thought I would be stretching my friend's hospitality if I stayed a third night, so I decided to spend the next night in the open.

I had modified an old army medical cot and put it in the bed of my pickup, which had a canopy. It was quite comfortable with a mattress and sleeping bag, once I took out several layers of assorted junk to make room for myself. But this could wait, I said, as I sat there that night looking up at the stars wondering… wondering about life on earth and elsewhere and reminiscing about my own existence, reminders of which leapt out of the glowing fire and hot embers. Somehow I thought back to the San Diego Nickel Snatcher when I was in the Navy in 1952. The Nickel Snatcher was a small personnel ferry for military and civilian workers traveling to and from the North Island Naval Air Station in San Diego Harbor. While waiting, I had bought a book by Frank Scully entitled *Behind the Flying Saucers*, and even though the light had been dim in the waiting area, I was enthralled. I had heard the coxswain reverse his motors as the boat glided into the dock. I was dimly aware of some sailors walking past me down the gangplank. I just wanted to finish the page. A loud voice shouted above the idling engine, "Hey mack, ya comin' or ain't ya?" As if to emphasize his point, he gunned the motor loudly. I turned down a corner of the page and just made it. As a young sailor with the rope jumped aboard, so did I.

As I looked into the fire, the past seemed to reflect through the orange light, all those earlier times came flooding back; The Navy years, college, the formation of Kansas City NICAP, working with Donald Keyhoe. Here I was almost half a century later staring into that fire, which had slowed time and focused my memories on past events; marriage to a wonderful gal, two kids, scout work, college, and 30 years as an educator, now only memories. Each memory was like an ember giving off sparks that disappeared into the blackness. There were a few more sparks now than there had been before. A cool wind was now blowing sharp quick gusts, scattering the sparks into the night air and seemingly into the stars above. I was glad I had learned in my scout days about building fires in open areas.

Out of my peripheral vision I saw a movement, but when I turned my head I saw nothing. Then I heard a slight rustling sound as the breeze picked up. This time my flashlight beam stabbed the darkness, just in time to catch a tumbleweed rolling a few yards away from camp. I pulled the sheepskin coat collar up and my hat down. The stars I had reveled at a short time before were now somewhat fuzzy and were beginning to be obscured by clouds. I scrunched down in my chair, determined not to let a little weather force me into the confinement of my small truck.

As I watched the sparks from the now-diminishing fire swirl around in the black vortex, I heard a ping as a large raindrop hit my coffee pot on the edge of the fire. Then splat! One or two of the huge drops hit me. Their coolness would have been refreshing if they had not come down with such force. Then more hitting the coffee pot. Ping! Ping! In the distance thunder rumbled. The once-dancing sparks were soon to become a smoking, sizzling black hole. I thought better of my determined stand against such odds. I started towards the truck a few yards away. The sudden storm did not give me time to crawl into the back of the truck where I had my sleeping bag. As I put my hand on the door, the landscape lit up momentarily like a bright silver light had been turned on. Out of the corner of my eye a jagged streak of lightning stabbed the ground and cracked the black sky from top to bottom. As I jumped in and shut the door, the lightning's crashing report reverberated in my ears. Then the sky lit up again, and I caught a glimpse of a gray sheet of rain coming even harder.

The windshield appeared to be under water, but there was some visibility out of the side windows. I shined my flashlight towards the fire. It was not even smoking anymore. The chair and the coffee pot were nowhere to be seen. Out of habit, I turned on the truck headlights. Great splatters of mud were thrown into the air, only to be knocked down again. It was a smashing onslaught. The din on the roof of the cab was deafening. Lightning flared again between bouts of intense darkness and pounding rain. I switched on the ignition and saw that I had over a half tank of gas, but I immediately switched it off. I was not going anywhere. The dampness of my clothing felt prickly and clammy. I managed to get my heavy coat off and manipulated my shoulder through the tiny rear cab window. I was able to get a warm sweatshirt on and my sleeping bag into the cab. I wrapped it around me; let it storm…I was warm and dry.

Somewhere in the wee hours I woke up and watched the moon play hide-and-seek behind some dark clouds. Here and there patches of stars had reappeared. I managed to hobble barefooted and somewhat gingerly to the back of the truck and crawl in. I had some difficulty arranging the stuff around me, but I dozed off.

The sun was over the mountains by the time I woke up, and it was going to be a beautiful day. As I crawled out over the tailgate into the fresh cool air, I looked up into the azure blue sky. No clouds were in sight. I managed to resurrect my aluminum chair from a piece of sagebrush, and my coffee pot from under the truck. I never did find the lid.

## WHAT IS THAT?

This was to be my last day in the arroyo basin, and I was going to thoroughly search it. Stashing my camera equipment in my pack and picking up my walking stick, I started out, walking back and forth across the basin, crossing the old braids of the old water courses again and again. The rain the night before had cleaned everything. I would stop here and there to poke something with my stick. I studied the various kinds of lichen, sat and rested, and started out again. I had just taken some pictures and was working back toward the gap when suddenly I saw something under some brush. It was totally different. It looked like a pile of solidified chicken fat above the sand. Part of it was buried. I carefully picked it up and brushed off a crust of sand on the bottom side. I did not know what I had, but after the publication of three books on pioneer history and tramping around old wagon and stagecoach roads, cabin sites, arroyos and pioneer junk piles for 25 years, I knew it was unique. Every piece of anything found can eventually be identified: bottles, car parts, barbed wire, tin cans, fruit jars, etc. Even the patterns on broken dish fragments help date old china.

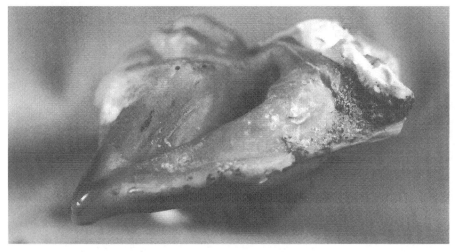

*Figure. 7.2. **Right side of artifact** as it was found. Dark flakes are congealed oil. Darker tip indicates the surface melted during intense heat. This right side may have provided some stiffness while the soft material functioned. Decompression or an explosion may have blown this artifact out of the crashing craft.*

This artifact fits absolutely none of these historical or modern identification criteria.

Two things immediately came to mind; 1) a petrified internal organ of some kind, probably belonging to one of the large animals that were in the arroyo basin (cattle, deer, perhaps antelope), 2) something from a Native American grave site, as there were many of these on the old terraces that were around the old lake bed, as reported by numerous teams of archaeologists over the years. I dared not even consider any other alternatives.

I spent the rest of the day criss-crossing the basin looking for anything else that might catch my eye. I saw nothing. I went back about mid-afternoon with a shovel and dug around the spot where I had found the artifact. I had hoped to turn up the broken-off pieces, but found nothing. I did note that the discovery site was higher than the main arroyo water course, and absolutely no water had flowed in the nearby water course in recent years because of a large blockage above.

As I criss-crossed the basin, signs here and there reminded me of the food chain. The prairie dog complexes had not been used in some time and most were collapsed in the center. A ranch wife had told me that a virus had wiped them out several years before. Once or twice I flushed a jackrabbit from the shade of some sagebrush. I saw the sobering rattle end of a rattlesnake disappearing into one old prairie dog hole. Some of the wildflowers were in bloom. I walked by numerous stands of white spectacle pod, saw a monarch butterfly on some purple Rocky Mountain bee weed, found stands of orange blazing star and the solitary white prickly poppy nodding in the breeze. The rain had renewed a symphony of color on the greening plain.

My father and mother had spent some of their early married life near Deming, in southern New Mexico. Dad was a mining engineer engaged in field exploration for new mining sites. Dad knew an author, Russ Calvin, who wrote a best-selling book called *The Sky Determines* published in 1934. This southwestern classic was aptly named. In the Plains country, a day may start without a cloud in the sky with visibility for 25 to 30 miles. In very flowery language Calvin describes how a land's history begins with what nature determines. This is certainly true on the Plains. As the day wears on and heat rises, small dust storms contribute to haze, limiting visibility. On a day when there will be "weather" by noon or so, huge tall clouds come in with high atmospheric winds. The day might be sunny and bright if the clouds keep moving or overcast with wind gusts or a short hard rain if they do not.

When I commented to a rancher about the huge raindrops I had experienced the night before, he replied, "Yeah...they get plenty big around here, but look at that sky they come from!" I had never thought of it in quite that way.

It was getting cooler; now a breeze had come up. Earlier some cumulus clouds had risen over the San Mateo Mountains. Now the sun slashed through them, casting shadows here and there across the ragged landscape. The shimmering heat of the midday had gradually given way to the long shadows and stark outlines of late afternoon. Periodically the sun went behind a cloud bank; when it emerged, the basin was bathed in a bright golden light. The sun would be below the mountains soon and I had better be on my way. I would stop at Datil for supper. It had been quite a day and I was going home tomorrow.

## OTHER FINDS

The author has researched and published books and articles

since the 1970s about the settlement of peoples who followed the demise and dispersion of Native Americans from 1890 to the 1940s. Like the Native Americans before them, the settlers left signs of their life around old buildings and especially in arroyos where much of their refuse ended up.

A smattering of 1920 to 1930s material was found throughout the area with concentrations in the previously mentioned arroyo braids. Most of the people in the immediate area of the arroyo are of Hispanic, working-class origin, and their refuse reflects this. They by and large are the descendants of Mexican vaqueros who were hired by wealthy homesteaders to work with cattle and horses in the early days before 1900.

There were some things laying around or found in small excavations that were anomalies and did not fit into the 1890s to 1940s time period of the area inhabitants. In addition to the larger and seemingly more important things found, the shoe sole, the artifact and smaller unfamiliar things came up. I kept only the odd looking things I could not identify. Some things had multiple parts. For instance, the wafer (Figure 7.6) had four or five materials that needed to be analyzed. Each item at the analytical chemist cost $50 to $100 to be identified. Identifying the materials was informative but still did not tell me what it was. The wafer had zinc-based tape around the perimeter, a

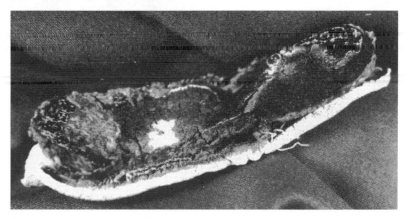

*Figure. 7.4. **Little shoe sole** is a big mystery. Measurements are 4 $^7/_8$ inches (12.3 cm) long and 1 $^1/_4$ inches (3.2 cm) wide. Sole is too small and well made for terrestrial adults, too long and narrow for a young child. (See Chapter 13.) Podiatrists tell us that shoes are sized according to the width of the ball of the foot or instep.*

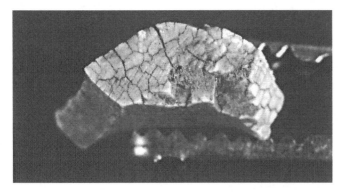

*Figure. 7.3. **Wax pieces** were found throughout the arroyo crash site. Some were thicker, like the one above, while others were thin. Most pieces had a curved outer side and a convex inner side. I got the impression it had gone around something.*

ceramic top, and charcoal interior. What was it? I finally had to accept I probably would never know. There were several other items that were anomalies to the above time period, which are briefly described in the remainder of this chapter.

## LITTLE SHOE SOLE

At the widest part, the shoe sole translates to nearly 5 inches long (12.5 cm) and 1.5 inches wide (4 cm) at the ball of the foot. Dr. Robert Leir, noted UFO implant researcher is a podiatrist. When he saw the shoe in 1998, he said, "I have worked on children's and adult foot sizes for years and have never seen a size this narrow."

One would think a shoe sole would be easy to identify. It was, but analysis of the layers was extensive. The analytical chemist indicated that seven of the shoe layers were cellulose from plant fiber. Research indicated that there were six main cellulose plant fibers: hemp, jute, flax, ramie, sisal, and cotton. I could guess but I had to be specific and that cost another analysis fee to have the right one identified. Who would have thought clay and calcium carbonate would be in a shoe sole, but it was. So was natural rubber, plant pitch, and urethane. It was somewhat disconcerting to find at one stage that I had a shoe sole with materials too expensive to be manufactured and sold, and in a size no one could wear. That's what I had, and I took several years to realize I had gone as far as I could.

## THE FABRIC SCRAPS

In the 2004 dig, we found many more metal foil shards. In addition to the shards, we found two pieces of cloth in the test holes dug in the main "gap." The pieces were tiny, very small-sized scraps. One piece was gray that we pulled out of some grass roots. It seemed interesting, with one side slightly different than the other. The other piece was embedded in a dirt clod. It was darker in color, and more coarse. When the fibers were sent to the forensic lab, they did not at all seem important. But since the lab is world famous for their fiber analysis, I decided to let them do it. This same lab was asked to analyze the fibers on the Shroud of Turin for the Vatican in the late 1970s. We knew they would do a first-class job, and we were not disappointed.

The samples were inspected with a stereo-microscope, and portions of each sample were prepared for microscopical study using polarized light microscopy (PLM). The lab uses infrared micro-spectroscopy (IR) to determine the composition,for scanning electron microscopy (SEM). The laboratory also studied the fiber morphology to determine elemental composition. Recalling reports that the gray aliens had tight-fitting, full-length garments, I asked them to check the fabric's elasticity and strength. They wrote:

> *Initial inspection of the gray material showed it to be soft, although not elastic, and constructed of a fabric-like material heavily coated with gray rubber.*

Major elements in the gray fabric were carbon, chlorine, silicon, aluminum, and oxygen. None of these are key components

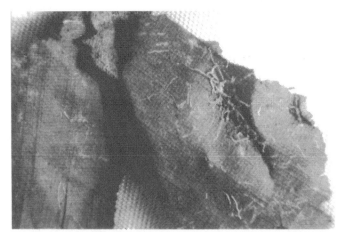
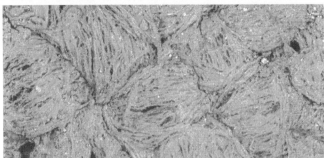
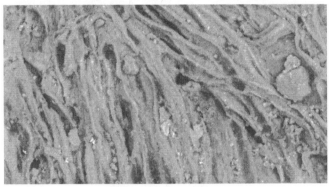

*Figure. 7.5. **Gray rubber coated fabric (top)** as found by the Navajo crew in 2004 with grass roots attached (10x magnification). **Rubber coated side** (second from the top, 100x magnification) note quilted pattern. (Third from top) **Rubber coating**, magnified 400x. **Black carbonized fibers**, use unknown, 18x magnification (bottom).*

of our rain gear. A rubber garment of any kind would be impractical for us to wear for any length of time with humidity present and body perspiration. Rubber is worn temporarily in many ways, including any area where moisture penetration would not be desirable. These include footwear, household gloves, rain gear, protective aprons, and diving suits. One of the most common uses of rubber today is medical latex gloves.

Rubber, of course, is a non-conductor of electrical current. We have read in UFO literature that the alien body is part of the energy circuit that guides the UFOs. Would this rubber-coated garment have been a part of this process? We have found flakes of wax we think might have been a protective coating inside the craft possibly to keep the aliens from coming in contact with metal surfaces. There was also rubber on the bottom of the little shoe sole. Is it possible the rubber acted as an additional insulator to keep the craft's occupants away from an energized surface? If the aliens were part of the circuitry of the craft, it is assumed they would not be wearing rubber-soled shoes when the craft was in operation. Was it possible that whatever energy the aliens may have been a part of flowed through their bodies, and the rubberized cloth acted as insulation to keep the energy channeled? Sort of like a rubber or plastic coating on a wire? One more question arises. Col. Corso said the army had discovered in their secret alien autopsies that the "EBEs (Extraterrestrial Biological Entities) had added artificial parts to genetically altered cloned replicas of themselves to better withstand the rigors of space."[1] If the aliens were part of the craft circuitry, would the HDPE artifact have improved or aided in this process? It is anybody's guess.

The black carbon fibers in Figure 7.5 (bottom photo) were given extensive testing and were found to be artificial. The lab said:

> **The black coated material was also studied using PLM and SEM. The fibers were macerated in xylene. The fibers appear fully carbonized. They now appear to be carbon fibers. Their striated surface is characteristic of carbon fibers produced from polyacrylonitrile (PAN) fibers. Regenerated cellulose fibers (rayon) also have a similar surface appearance... The type of fiber was not further investigated.**[2]

I might add that these fibers were also found in what we have termed the "explosion zone." The approximately 20-yard diameter area where many of the other smaller artifacts were found.

## A CUP OF COFFEE AND A ZINC WAFER

As I was leaving the Plains on one of the late 1990s trips, I stopped by to say goodbye to a ranch wife with whom I had become friends. The aroma of her kitchen was always great, and something good was usually simmering on the stove. Today it was clam chowder, so thick and good, a spoon could stand up in it. Naturally, I gave in when my arm was twisted to stay for a bite. This bite was an understatement; it included a huge hot bowl of chowder, a thick slice of homemade dark bread, well-buttered, a huge glass of iced tea, and an ample slice of apple pie with hot strong coffee. Ranchers like their coffee strong, and this gal upheld the tradition well. So well, in fact, that the boiled coffee sitting on the back of her wood stove in the inviting blue

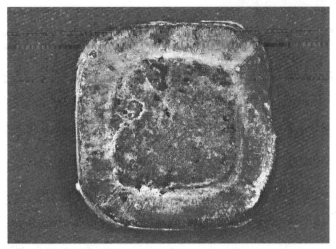

*Figure. 7.6.* **Back of wafer** *has radial lines around the perimeter. It is 1 inch (2.5 cm) square and $3/16$ inch (0.5 cm) thick. The outer front covering is some sort of ceramic material and the backside seems to be charcoal.*

enamel pot, had to be poured into the cup through a tea strainer to remove the grounds. The western coffee tradition still dictates its coffee campfire style, even if some coffee grains still manage to find their way into the cup.

Along towards the second cup, she put a square-like item on the table and said her husband had found it in the arroyo and wanted me to have it. It was about the size of a square snack cracker, but twice as thick. I picked it up, got out my magnification glass, and looked it over. The top had a ceramic appearance and was slightly caved in. Layers were visible in the center; the back appeared charred with some pits in it. It had a concentric ring around the outside, which reminded me of the wiring of an electric armature. The four edges were all different indicating the construction of the wafer, as I came to call it, as deliberate and planned. One edge had some fibers showing, but without a stronger magnifying glass I could really do nothing with it until I could get home and have a FT-IR spectroscopy run on it. Additional research is in Chapter 13.

## CHAPTER 7
### REFERENCES

1  Kent, Bill and Hester. "Designer Aliens"; *UFO Magazine*, Vol. 17, No. 1, pp. 42–43.
2  Campbell, Art. Correspondence and Lab Report from McCrone Associates, Chicago, p. 1.

# CHAPTER 8

## THE GRADY BARNETTS IN 1947

In the 1980s I had read that a civil engineer with the Soil Conservation Service had allegedly discovered a crashed flying saucer in 1947 up on the Plains of San Augustin (*The Roswell Incident*, Berlitz and Moore, 1980).[1] Since the story was not mine and I thought it had been amply covered by other authors, I really had no plans to pursue it. That was until a chance meeting at Harold's Gift Shop in Socorro, New Mexico.

After I had finished taking the soil samples at the site in the summer of 1997, I drove to Socorro to return a borrowed government Geiger counter and a rented post-hole digger. I decided to stop at an attractive curio shop on Highway 85 to buy a present for my wife. As I walked into the shop I heard soft native American music and a tinkling of some small wind chimes. While I was browsing, a pleasant woman asked if she could help me. A conversation started that gravitated to where I was from and my work on the Plains and the UFO research.

I was surprised to learn that she and her husband had been neighbors and friends of the civil engineer named Grady L. (Barney) Barnett, who claimed to have come upon the crash in 1947. She pointed one block away and said, "We lived across the street from them over there on Park." She thought her husband, who would return soon, might like to talk to me about it. I learned that their names were Martha and Harold Baca. I agreed to return in an hour or so, which I did.

When I returned Harold was there, and while his wife assisted customers, we retired to a lower part of the shop, which held colorful boutique items, shelved ceramics and native American blankets. Harold Baca was a man who had been around a half a century or more. He had been active in several organizations including a muzzle loaders association, Socorro Lions Club, and the local American Legion post. He had served in the U.S. Army and had been stationed in Germany. Many men of the Southwest are hunters, but Harold was a special breed. He had done much of his hunting with bow and arrow. At the times we chatted, he was a licensed general contractor, member of the Presbyterian Church, and a board member of the Socorro Electric Co-op.

When we started talking about his neighbor Barney, his voice lowered a little toward reverence. He referred to a time the summer before Barney's death, when Barney was sitting outside one summer evening in his favorite chair. Harold, gesturing with three large fingers to his throat, motioned me closer. "That's what Barney did the day he told me about the little people." He went on to say, "Barney had throat and lung cancer, you know," he said pointing to his throat. "He couldn't speak too loud." He related that Barney told him, "The little people did this to me, the day I bent over them on the Plains of San Augustin. They were radioactive and they did this to me."

The Baca family had been in Socorro for many years and is well respected. Harold and his wife Martha had moved across the street from Barney and Ruth in 1966, but they knew them only as neighbors and little about their private affairs. At a later time, after Barney had told his neighbor about "The little People," Harold asked Ruth, "Was Barney hallucinating about the story or what?" Ruth said, "Oh no! It really happened out there on the Plains about twenty years ago. But we just do not talk about it much anymore."

Harold had one closing comment. He said, "We had other things to do at that time. We were raising a family of four and I just didn't discuss it anymore with them. It wasn't until later on that I grasped what the man had seen. He died and I regretted not getting more information."

*Figure 8.1. **The Barnetts** in 1948 about a year after Barney's experience. Ruth kept a detailed diary during 1947, but no mention is made there of a flying saucer crash. They told friends and relatives, but agreed not to write about it.*

Over the years, Barney told his story of the UFO crash to nine or ten people that we know of, and six of those are included in this narrative. They include Verne and Jean Maltais, Harold Baca, William Leed, and Ruth Barnett's niece Alice Knight and her husband J.B. Alice was kind enough to loan us a copy of the diary Ruth Barnett had written in 1947. There is no mention of the UFO crash in the diary, but it includes all of Ruth's and Barney's social activities. In the front of the diary Ruth listed her name, their addresses at 520 Park Street, and the phone number at the time, "152 M." It also contains details about their new house under construction next door and monies spent. Ruth almost always recorded Barney's movements to and from his work — if he was in the office or in the field and the times he returned. I discovered little was known about his private life, his youth, education, and his service to his country in World War I, the geographical area where he worked, and specifically what he did.

The veracity of any story begins with the credibility of the witness. This is often tied to the observer's lifestyle, friends, associates, and integrity. What is unique about Barney's story is that Ruth's diary covers only the year 1947, a significant year in Barney's life. We learn about his daily life, his travels, his social life, and even his relationship with his wife Ruth in their thirtieth year of marriage. Their daily routine is almost like clockwork. Barney left for work about 7:45 a.m. He would drive their car or Ruth would take him if she needed to run errands. Her day was filled with housework, laundry, shopping, ironing, cooking, and visits with friends. When in town, Barney came home for lunch. Although the main Soil Conservation Service office was 26 miles away in Magdalena, Barney had the use of a desk and drafting table at the nearby Highway Department office. He was usually home by 5 p.m.

Ruth and Barney lived an extremely frugal life. They owned no washing machine or dryer, as indicated in the diary passage, so Ruth took their clothes to the Helpy-Selfy Laundromat. To save money she would bring the clothes home damp and hang them on their own clothes line. Doing her clothes like this cost less than one dollar per week. They lived in a rented house. Rent was $6.00 per month. Ruth wrote, in the fall of 1947, "Mosquitoes terrible, Barny kills some every night." They heated their house with wood in their fireplace which cost $10 to $12 a ton. It took about one ton of firewood to heat the house in the winter. Civil service salaries in those days for engineers started around $1,600 per year. Barney with ten years experience probably took home less then $225 a month. Ruth cooked over a coal stove in the kitchen, winter and summer, and canned in the summer to augment their food supply. She seldom spent money on herself; and the diary lists only a few purchases she made for herself, one house dress for less than $10, a pair of popular "wedgie" shoes and a summer handbag. [2]

Alice Knight said her aunt Ruth, when she was young, had beautiful long, very dark hair. Even as she became older she wore it long and put it up in a stylish way. Ruth's one excess, if it could be called that, was to have her hair shampooed at Letha's Beauty Salon once or twice a year or on special occasions. Cost, $1.25. Barney, on the other hand, became relatively bald by age 40 in the early 1930s. Pictures of him in WWI and the early 1920s show him with a very receding hairline. Ruth used to joke that she had enough hair for both of them. I asked Vern Maltais if Barney ever wore a western hat. He replied, "I never saw him in one. He never considered himself a rancher or cowboy type. He was proud to be an engineer and only wore an old fedora outdoors to keep the sun off."

They did have a lot of expense going towards a new house they were building next door. It was a basic slab floor design with a stucco exterior. Barney hired a man named Lopez for $6 a day to work on the house, and he joined him at every opportunity. Barney and Ruth made eight or nine trips to Albuquerque for materials. At least two loans were taken out in 1946 and 1947 for materials and labor costs for the house. Ruth was thrilled with several items for the new home and kitchen. She listed a Hoover sweeper, a two-burner hot plate, a Presto cooker, a dishwasher, and venetian blinds. Ruth got busy and planted flowers, rose bushes, shrubs, and a tree or two in the new yard. The house was almost finished in early December of 1947 when they moved in.

The Barnetts led an active social life. They usually spent one weekend evening with friends and sometimes they went to the Lowe Theatre for a movie. Ruth belonged to the Westminster Presbyterian Ladies Guild, which met in members' homes and held several fundraisers each year. She met almost daily with friends who would drop over or she would go to see them. They quilted, exchanged magazines and recipes, chit-chatted and sometimes went to Miss Betty's for coffee or cold drinks. Barney went almost without fail to a dinner meeting of the Socorro Rotary Club every Monday night. He was a member for some seventeen years until his health began to deteriorate about 1965. They also attended functions at the New Mexico School of Mines. In a small town like Socorro, engineers were a rather select group and they were invited to the college activities, including lectures, concerts, and films. Once in 1947 they took the newly married Alice Knight and her husband J.B. to a prom at the college.

*Figure 8.2. **Harold Baca** was a neighbor of the Barnetts. He was told of Barney's Plains experience in 1968, the summer before Barney's death. He may have been the last one Barney told. Locally, the Baca name is greatly respected. Harold and his wife run a curio shop on Highway 85 in Socorro.*

Their evenings were almost always spent together quietly. Alice Knight said that Ruth "adored Barney" and their evening life reflected a mutual respect. After the supper dishes were done in the winter months, they would sit by the fire and listen to favorite radio programs. They were avid readers and spent hardly anything on reading material. Their friends in the various towns they had lived in would send them old newspapers. Ruth's diary mentions receiving bundles of the *Dalhart Texan* several times during the year. They also exchanged magazines among their many friends including the *Ladies Home Journal*, *The Saturday Evening Post*, and the *Readers Digest*. For

Christmas every year Ruth gave Barney a subscription to a monthly magazine for amateur astronomers called *The Sky* which cost $1.50 per year. In the evenings they would return correspondence, take baths, and Ruth would comb her hair and put it into a soft braid for the night. The last event of the evening was the 9:30 news which was usually followed by bed. As the days got longer, they worked on their new house and yard in the early evening and on weekends.

In 1947 Ruth and Barney had some health problems. They shared several colds and the flu. There were other entries, mostly about Barney's health. He stayed home from work four or five days that winter and had a throat infection and some blood pressure problems. Alice Knight said he had a, "Bad heart attack in 1949" and, "He had a swallowing problem after World War I, as he was gassed, and spent a lot of time in a hospital immediately after."

**THE EARLY YEARS**
Barney was born in Amarillo, Texas, on October 22, 1892. His father was Ike Barnett and his mother was Katie Benson. Barney was the oldest of three children, having two younger sisters. The family lived in the Goodnight, Texas area. It is believed Barney started school there in 1899. Schools in those years were of the one-room variety, and education usually terminated in the eighth grade. Barney finished his initial schooling about 1907. The year Barney was born a college was formed near his home. In 1905 it became the Goodnight Baptist College, which was known locally as the Academy. A photo dating from 1909 shows Barney on the 1909 football team. It is believed he was a key running back and team captain.

Barney attended the Academy from 1908 until 1912. After graduation from this local school, he attended Humboldt College at Humboldt, Iowa, where he had a role in a 1913 commencement play. By the end of the 1914-1915 school year, Barney had earned a Bachelor of Science degree and was named valedictorian. In the commencement program he was listed as being from Amarillo, Texas.[3] The school year of 1915-1916 found Barney on the west coast at the Oakland, California, Polytechnic College of Engineering completing his requirements for civil engineering.

After leaving college in 1916, Barney worked as a civil engineer in Boise City, Oklahoma and Stockton, California, where he did surveying, topographical mapping, dam construction, and irrigation work. Barney also worked for the Navajo Cooper Company of Flagstaff, Arizona, in early 1917.[4] Here he put his engineering skills to work. Barney had known Ruth for some time, and in 1915 they started seeing each other. When Barney returned

*Figure 8.3. **Barney and Ruth Barnett** shortly after their marriage in April of 1918. Car is a 1917 Dodge. Barney's leadership and professional abilities helped him rise from enlistee to captain by his discharge in 1919. Photo courtesy of Alice Knight. Barney, in an engineering unit, rose from an enlistee in 1917 to Captain in 1919 when he was discharged.*

to Oklahoma from Flagstaff in the fall of 1917, it appeared he would be going into the Army. President Woodrow Wilson had asked Congress in April of 1917 for a declaration of war with Germany, and trained engineers were very much needed.

Barney and Ruth were married on October 3, 1917.[5] He enlisted in the U.S. Army on December 21st of that year and was sent to Camp Travis, San Antonio, Texas, assigned to the 47th Company, 12th Battalion of the Depot Brigade. In January of 1918, he was continuing his training at Camp Owen Biernie at Fort Bliss, Texas. He was a supply sergeant in Company C of the 15th Cavalry Division.[6]

Barney received his appointment to officers' training school in March 1918 and was sent for training to Petersburg, Virginia. There he was commissioned as a Second Lieutenant in the Corps of Engineers on May 19, 1918.[7] To many of the men of the 88th Division to which Barney was assigned, it appeared that the 88th would be a training division to which men were assigned, trained and transferred to other units. The division was in a huge encampment at Camp Dodge, Iowa. The division included these units:

- Division headquarters
- Three machine gun battalions
- Six infantry units
- The 313th Engineer Unit (to which Barney was assigned)
- Several special training units
- A medical unit

The division strength was huge, about 28,000 men. Today Barney's unit would be considered combat engineers, as they functioned inside a combat unit and had the military training and weapons to defend themselves in wartime. On the big push the allies were planning just before the Armistice, some of the 313th were slated to be used as ground troops, but the war ended before these plans could be carried out.[8]

In the fall of 1917, some officers from the British and French Armies arrived at Camp Dodge to help with the training of the 88th Division. The training was in grenade warfare, bayonet combat, automatic weapons, field fortifications, sniping, camouflage, use of trench motors, signal liaison, and gas defense.

Barney and some of the other officers of the division started to think about sending for their wives. Barney told Ruth in one letter written in June 1918 that he was looking for a place for them to live so she could join him the next month. Before their plans could be made, the Unit received orders to go overseas[9]. Travel on the seas still had its dangers. Earlier in 1918, 113 forestry engineers trained in Washington State had drowned when their troop ship, the converted British liner *Tuscania* was torpedoed.

*Figure. 84. **Barney Barnett** at Camp Dodge, Iowa, 1918. He was commissioned a 2nd Lieutenant in the 313th Engineers and assigned to the 88th Division. His division was on the front line a short time and experienced one gas attack. Photo courtesy of Alice Knight.*

By August Ruth had received letters from Liverpool and Southampton, England, and the embarkation port of LeHavre, France. Later some letters were received from a training area near Semur, France.[10]

No sooner had they settled into their training routine when the 88th Division received orders to go to the front and relieve the American 29th Division. It was found that not all the 88th helmets and gas masks had been shipped with them and the division was sent to a quieter rear sector to await them, which delayed their going into the line until October 12th, near Belfort, France. The 88th was assigned to a wooded, rolling plains area just inside Germany known as the Belfort Gap near the Vosges Mountains. The main body of troops were put in the line with the 38th French Division. The plan was that after a ten-day orientation the French would withdraw. Much of the replacement movement was carried out at night with a good deal of secrecy, as the Germans were maintaining a close aerial observation of that sector at that time.[11]

Facing the 88th were divisions of the German Army detachment "B" and the 30th Bavarian Reserve Division. The line the French vacated to the Americans was some 17 kilometers long, and the area (No Man's Land) between the Germans and the Americans varied from 1 kilometer to 300 meters wide. Some of the French trenches were in poor shape. Some were partially filled with stagnant water and some of the trenches were caved in. The 313th was assigned to rebuild and make stronger important parts of the trench system. Companies of the 88th Division were also assigned to make these trenches more habitable for the men.

On the night of October 12-13, 1918, near the center of the line close to the French town of Blaschwiller, two parties of the 88th Division were sent out to reconnoiter a German trench.[12] The parties were cut off by a German minewerfer barrage (a muzzle-loaded grenade-throwing weapon). Many of the Americans were captured and several killed. As if to welcome the 88th, the Germans gave the French and American positions a barrage of high explosive fosgene gas shells. It was believed that Barney and some of his men were exposed to some of this gas, and he had health problems for years after his discharge.

Early in November of 1918 the 88th was assigned to a newly organized second Army. An attack was begun on the German lines on November 10th. The lead divisions were the American 7th, the 28th, the 33rd and the 92nd Division of African-American troops. The 88th was in reserve. The exhausted Germans fell back along a 50 kilometer front. The next day on the eleventh hour of the eleventh day, the Armistice commenced.[13]

In a letter posted from Nancy, France on November 22, 1918, Barney wrote from a Red Cross hotel reserved for officers, "Leave tomorrow to go to Paris. This is a large place and has been air-raided more than any other city in France. A large bomb hit this hotel (a large four-story stone house). The bomb went through the roof and buried itself in the concrete floor in the basement and did not explode!"[14]

We do not have specific data on casualties of the 313th, but during the forty-three days they were in front line areas near Belfort, France, the 28,000-man 88th Division records twenty dead and fifty-eight wounded. After the Armistice in 1919 Barney's Company C of over 280 men were ordered to the port of La Havre to board the *U.S.S. Buford*. It was a slow-moving ship, and Barney wrote to Ruth that it had carried troops in the Spanish-American War, some twenty years before. In another letter written at sea on April 18, 1919, he said, "Rough going through the Sea of Biscay, seasick." The *Buford* finally docked in the U.S. Most of the rest of Barney's unit and division came home a month or so later on the *U.S.S. Madawski*.[15]

Barney lamented to Ruth that he had been promoted to captain since the unit went into the line in October of 1918, but the paper work had not caught up with him before he was discharged. Before he left the Army, he had one last assignment - escorting 50 men to the Washington Barracks in Washington D.C., where he applied for an immediate discharge and left the Army on May 27, 1919. He and Ruth were reunited a few days later. Barney was twenty-seven and Ruth was twenty-four.

After the war, Barney and Ruth found themselves in the northwest Texas panhandle at Stratford, where Barney got his first county engineering job, working for Sherman County.[16] He and Ruth lived at Stratford between 1919 and 1921. Barney had several jobs in the 1920s. He worked for the J.I. Case Tractor and Motor Company in Amarillo, Texas. He sold farm equipment, as well as Chevrolet cars and trucks in Dalhart, Texas, and he was a park ranger at Ute Park, near Logan, New Mexico, where he and Ruth managed a lodge. When the Depression came along, Barney and Ruth moved back to one of the farms his father owned near Boise City, Oklahoma, until, as Alice Knight said, "The Dust Bowl starved them out."

Barney and Ruth then moved into Boise City, where Barney was out of work, although he did manage to work on and off for the Panhandle Radio & Electric Company.

It is believed that while working for the electric company, Barney turned his engineering skills to invention. He developed a device called the Magnetic Balance that located metal objects underground. He probably did not have the resources to develop it, but it did catch the attention of a Dr. H.H. Nininger, who was with the Colorado Museum of Natural History. Dr. Nininger field-tested the device at a place known in Texas as the Odessa Crater. He uncovered twenty-seven meteorites ranging in weight from one ounce to four pounds. The device was somewhat cumbersome, but it worked, and Barney's device was given a feature article in the February 1939 issue of *The Sky – Magazine of Cosmic News*.[17] He was to work on this device up until the middle of World War II, but never pursued it beyond the hobby stage. According to a 1948 *Socorro Chieftain*, Barney demonstrated his device, which was now called "The Doodle Bug," to the Socorro Rotary club. In the only known quote of any kind that was published, Barney read from *The Sky*, "It (the device) will spot gold and silver deposits, but that would call for a lot of digging, and he has not gone after that sort of stuff."[18]

*Figure 8.5. **Barney Barnett**, 1892-1969. Barney was valedictorian of his Humboldt, Iowa college class in 1914 as well as captain and running back on the football team. He attended Oakland, California's Polytechnic College of Engineering in 1914-16. Photo, ca. 1922, courtesy of Alice Knight.*

### ENGINEERING

Between 1935 and 1937 Barney was the county engineer for Cimmaron County, Oklahoma. In 1937 he went to work for the U.S. Department of Agriculture's Soil Conservation Service (SCS) in the Baca District at Springfield, Colorado. He worked briefly for the SCS in Clayton, New Mexico, before moving to the Mesa SCS in Mosquero, New Mexico, in 1939.[19] Barney was to remain in Mosquero during World War II. While there, he joined the American Legion and was elected commander of the unit sometime in 1940 or 1941.

During World War II, there were numerous war bond drives, and the one at Mosquero was one of the largest patriotic events ever held in that area. There was a rally and speeches, a parade and all kinds of fundraising to "aid our boys overseas." Barney was familiar with an Army Air Force base near where he had grown up in Dalhart, Texas. Barney contacted the base commander about sending up a military contingent for the bond drive effort.

The proceeding says nothing about UFOs and the WWII effort, but much about character. It was 1943, and the air base sent two jeeps of men to Mosquero. Master Sergeant Verne A. Maltais was in charge of the unit. Master Seargant Maltais served in the U.S. Air Force from 1939 to 1945. He earned his stripes the hard way. When the Japanese invaded the Dutch East Indies in 1942, Maltais and two other military men refused to surrender. They "acquired" a sailboat and took it through the Japanese waters 350 miles to Australia. They posed as native fisherman. Verne said, "When the Japanese planes buzzed us, we had darkened our skin with coal soot and waived at them. Sometimes, they waived back." *Photo page 67.*

After meeting Barney and his wife in 1943, the Maltais's became good friends. He told the author they were invited to Barnett's home several times.

One of the activities the new friends enjoyed together was looking for Indian artifacts. In a letter to the author, Maltais recalled that "Barney had sent a collection of artifacts to Washington, D.C. He collected arrowheads in the Dust Bowl of the thirties. He was an expert in this area." Apparently, with the top soil blowing away throughout the Southwest, those years many Indian artifacts were exposed. Maltais' letter told the author that Barney had sent five-thousand or so to the Smithsonian Institution.

During the research on Barney's background, I was able to acquire one of Barney's civil service commission applications from 1941. Barney listed a variety of engineering experience when he was a county engineer in Texas and Oklahoma. They include: designs of highways, bridges, supervision of all county WPA projects, submitting plans of state offices for approval, working with an architect in designing high school buildings, making topographical maps, and supervising survey crews. Barney designed and supervised construction of six Oklahoma dam projects from 1935 to 1937. The projects included the Squaw Creek Dam for the Oklahoma State Fish and Game Commission, the Carrupah River Dam, the Wilson Dam, the James Dam, and the Kenton Dam on the Cimmaron River near Kenton, Oklahoma.[20]

Barney had one other area of expertise that greatly aided him in his work. He had knowledge and experience with heavy equipment of the day. He also mentions that he had direct supervision of all county equipment including Caterpillar tractors, road graders, a pile driver, and other bridge-building apparatus. In one SCS district, he worked for, he had direct supervision of all motor vehicles.

The Salado Soil Conservation District (SCD) was formed in Magdalena, New Mexico, just before the outbreak of World War II. After the war Barney transferred to the Salado District in 1946. The home office of the Salado District was in Magdalena, some twenty-six miles west of Socorro. As SCS men did in those days, he chose to live in the nearest large town, which was Socorro. The eastern boundary of the Salado SCS district was about one mile west of Socorro.

Barney spent about half of this time at the Socorro SCS office where he had engineering services necessary to his work. He could consult with drafting and blueprint facilities for his map work. The other half of this time was spent in the field. Also, he had a desk at the Salado office in Magdalena. It is believed he was in this office once or twice a week, meeting with Salado personnel, receiving and answering mail, going over agreements, and possibly meeting with land owners concerning work to be done or in progress.

From information we were able to gather, he was an extremely competent engineer. His engineering specialty was surveying. He was encouraged to continue in this area by instructors at Oakland Polytechnic. He was quiet, efficient and well respected by friends, neighbors, professional peers and fellow Rotarians. One of those Rotarians was Holm Bursom Jr., a Socorro banker, who was elected in 1948 as Socorro's mayor. Interviewed in 1979 by Stanton Friedman, he said that he had known Barnett quite well. Bursom spoke highly of Barney.[21] While in Socorro in 1998, I made contact with the Socorro Rotary Club. Barney was remembered fondly by some old-timers, but apparently, held no Rotary offices. The Rotary motto is "service above self." Seldom do we find citizens of organizations like Rotary involved in controversy or hoaxes.

*Figure 8.6* **Ruth Barnett**, *1895-1976. She was Barney's devoted wife for almost 60 years. Originally from Indiana, her maiden name was Kelp. She met Barney in Stratford, Texas about 1915. Alice Knight recalls her beautiful long hair which she seldom cut. Photo, ca. 1922, courtesy of Alice Knight.*

Rotary is not a service club that anyone can join. Its membership is composed of community leaders and upstanding individuals of good stature and community respect. One may be asked to join only if the members consider the individual to be of high moral character and a contributor to the progress and well being of the community. Only a few are chosen from each professional field.

We asked Alice Knight what she remembered most about her Uncle Barney. She had known him as she grew up, as Ruth used to baby-sit while Alice's mother, who was a widow, was working. She said, "He was a jolly fellow. All of the nieces and nephews adored him." Barney not only had knowledge of geology but astronomy also. Barney apparently read a magazine for amateur astronomers called *The Sky-Magazine of Cosmic News,* as previously mentioned. Alice said, "He took me to the mountains and told how they came into being. He would sit outside in the evening and tell us all about the stars." [22] Barney must have wondered, as many inquisitive minds would have and do today, about intelligent life in the universe. Little did he know that someday he would be privy to the answer to one of mankinds' greatest questions.

Harold Baca may have been the last person he told about "the little people on the plains." It must have initially been a considerable burden to himself and Ruth to have knowledge of such an event. After Barney's disclosure to Fleck Danley and a rather rude rebuff, he no doubt thought it was in his and Ruth's best interest to remain silent about the event. In February of 1950, he told Verne and Jean Maltais, who were driving through and stayed a few days. Vern was a retired U.S. Air Force Master Sergeant. We think Barnett may have sought his advice about dealing with the military. Juanita and Walter Ames were good friends of Barney and Ruth. They were living in Socorro when Barney had his experience in 1947. He may have confided the event to them as Ruth's diary indicates they spent a lot of time together after July 3rd and into the fall months. With the exception of possibly these friends and relatives, Barney was, for the most part, to stoically carry his burden with great dignity the remaining two decades of his life.

# CHAPTER 8
## REFERENCES

1. Berlitz, Charles & William M. Moore, *The Roswell Incident*, Grosset & Dunlap Publishers, New York, NY, 1980, p.53.

2. Barnett, Ruth; Diary for the year 1947; Entry from March 20, 1947.

3. Humbolt College Commencement Program; June 10, 1914; Humbolt, IA.

4. Barnett, Grady L.; *Application for transfer, temporary appointment reinstatement or promotion*; United States Civil Service Commission, Washington, DC from Mesa Soil and Water Conservation District, Roy, NM to Salada Soil and Water Conservation District, Socorro, NM; January 1946.

5. Campbell, Art. Correspondence with Alice Knight.

6. National Personnel Records Center, (Military Personnel Records) St. Louis, MO.

7. Ibid.

8. *The 88th Division in the World War, 1914-1918 (unit history #04-88-1919)*; US Army Military History Institute, Carlisle PA; Printer Wincoop Hallenbeck Crawford Co., 80 Lafayette St., New York, NY; p. 20.

9. Ibid.

10. Knight, Alice; Personal correspondence with Art Campbell concerning Barnett's marriage and World War I military service

11. Op. cit. *88th Div.*

12. Ibid.

13. Ibid.

14. Knight, op. cit..

15. Ibid.

16. Op cit. Barnett's *Application for Transfer.*

17. Nininger, D.H., Dr.; *Sky Magazine of Cosmic News*; vol. 3, No. 4; American Museum of Natural History; 1939; p. 6, 7, 23

18. Ibid.

19. Op cit. Barnett's *Application for Transfer..*

20. Ibid.

21. Campbell, Art. Contact with members of the Socorro Rotary Club.

22. Knight, op. cit.

# CHAPTER 9

## BARNEY'S WORK, TRAVELS & DISCOVERY

If we are going to assess anything about Barney Barnett's life in 1947, it would be advisable to understand something about the organization he was a part of, what he did, and where he worked. Barney and Ruth arrived in Socorro, New Mexico, sometime in 1946. Barney had transferred there from another much smaller Soil Conservation District northeast of Socorro. The Soil Conservation districts in New Mexico were authorized by the state legislature on March 17, 1937.[1] The Salado Conservation District where Barney worked was formed by landowners living in the district by elective referendum on October 14, 1941.[2] It officially became a public body eleven days after Pearl Harbor on December 18, 1941. The Salado name originally came from the Spanish word for salt, brackish, or briny, and is the name of a large dry river and drainage area in northern Socorro County. It runs through the northeast corner of the District, and is a major drainage into the Rio Grande River some 25 miles north of Socorro.

The Salado District has about one-half of its land in Socorro County and the other half in Catron County to the west. The District also extends about one mile into Valencia County, on its northernmost border. In the 1940s and 1950s the Salado District included an area 72 miles wide east to west and 70 miles long north to south, some 5,040 square miles. See map, (Figure 9.3). The District's eastern border begins about one mile west of Socorro where it crosses U.S. Highway 60. It goes north to the San Lorenzo arroyo where it begins a series of steps, every six miles or so, along township lines north into Valencia County. South of U.S. 60, the eastern edge of the District works its way around the Cibola National Forest boundary to the original Salado southern boundary.

The legal announcement for the election to form the Salado District was in the *Socorro Chieftain* on September 25, 1941. The legal description of the district specifically mentions the outside boundary of the Augustin Plains Ranches.[3]

*Figure 9.1 Fleck Danley, 1911-1999, sits astride a horse named Rooster. Danley was Barney Barnett's boss at the Salado Soil and Conservation District and a rancher in the Magdalena area after WWII. He learned of Barnett's Plains crash experience the day it happened. It is believed Fleck consulted with his superiors at the U.S. Department of Agriculture about Barney's discovery. He was concerned about government take-over of Plains range land. Fleck (his given name) served over 30 years on the SCS Board. Photo courtesy of Beth Danley.*

The soil conservation movement grew out of the dust bowl days when millions and millions of acres of agricultural topsoil in the southwest simply blew away. The parent organization of the Soil Conservation District was the United States Department of Agriculture. The Soil Conservation Service Work Unit #2, as it was known, was listed in the 1947 Magdalena-Socorro phone book as telephone #1.[4] The SCS assigned agricultural experts to work with landowners. The main function of the SCS was to provide technical assistance without cost to the farmers and ranchers. To receive the assistance, local landowners needed to form into districts, elect a board and set priorities. The Salado District had three types of areas that needed management. They were the Salado watershed, the mountain areas, and the Plains of San Augustin. Almost all of the work done in the Plains dealt with range management and grazing areas. Much of this effort concerned retaining rainfall, retarding soil losses, and preventing erosion in the water courses, many of which led into the Plains.

The following is a brief sketch of Barney's boss, Fleck Danley. Fleck was on the Salado Soil and Water Conservation Board whose office was in Magdalena, 26 miles west of Socorro. He worked closely with Barney, who was the designated district engineer. In 1947, Fleck was thirty-seven years old and had just returned from Germany after serving in the Army for four years.

Fleck was one of seven children and spent his schooling years in Alamogordo, New Mexico. After WWII, he and his wife, Beth, began ranching near Magdalena. Fleck was to serve on the SWC board for over thirty years. Although Fleck was a small man, his name did not reflect his size, but happened to be his given middle name, his first name being James. He was a rancher all his life. Those who knew him said that he was a good businessman and second to none when it came to horses.[5]

Through their local districts, the landowners signed cooperative agreements for specific projects. Once the agreements were signed, technical assistance would usually follow. A farmer in the Socorro SWCD might need different advice on improving his cantaloupe crop than a ranch owner near Magdalena running cattle. Each specific district had professionals who were hired through the U.S. Department of Agriculture to meet the needs of the landowners.[6]

The Salado District signed fifteen cooperative agreements during 1947 and included some 318,468 acres.[7] The map (opposite) indicates where these ranches were located and their proximity to various parts of the Plains.

The 1947 projects included stock water development, which involved placing a steel water tank or a small pond near a water source. Contour range furrows were designed and constructed. These were horizontal ditches built on a contour to help prevent erosion. The District also signed agreements to put in dry land farm terraces, which are horizontal ditches to drain water.[8] The costs for these projects were often paid for by the landowner, although there were some federal monies available for specific types of projects. Most of the work was done by independent contractors who had their own equipment and were hired by the rancher. Some of the ranchers we talked to appreciated the technical assistance from the Conservation District, but would rather proceed on their own and keep government involvement in their lives to a minimum. Much of the land on the Plains is owned by the state of New Mexico and the U.S. Government and is leased back to the ranchers for grazing.

## BARNEY'S WORK AND TRAVELS

Barney's role in all of these activities entailed a lot of surveying work. Since the range land was usually not clearly marked on the ground, he had to establish property lines to outline the parameters of a project. In land surveying, the survey needs to establish township, range, and section corners on a tract of land. A township is six miles square, containing thirty-six mile-square sections. A surveyor works from known previously surveyed points. He needs to find a direction or compass bearing from these known points to re-establish markers or section corners. Sometimes the known points are lost or otherwise obliterated, and sometimes they are some distance away from the property to be surveyed. Once the known points are established, the surveyor measures the distance between corners in feet. The direction is measured in degrees, minutes, and seconds. All of these measurements must be established by a highly sophisticated surveyor's compass and sighted by a transit in the hands of a competent surveyor.

In addition to surveying, Barney did a lot of project inspection on works in progress or those newly finished. In laying out an irrigation ditch or a dry land farm terrace,[9] the transit and survey rod must be set up to find the proper grade for a ditch, which should drop one to two inches per one hundred feet. Usually Barney had a back chainman to hold the rod while he was some distance away with the transit. This may have been someone associated with the ranch or someone hired to help him. On one November occasion, Ruth went along with him and, "held the rod for seven or eight shots," as she mentioned in her diary.[10] He also supervised field leveling, which is necessary to plant range grass. He made maps of his work and did monthly reports.

In order to find known points for his land surveying, Barney would need to occasionally visit the county surveyor's office to look at survey records. The county seat of Socorro County was the town of Socorro. This would not have been much of a problem because he lived there, but if he needed to know about land records in the western half of his District he needed to go to Reserve, the county seat of Catron County, some one 129 miles away and 40 miles south of Salado District's southern boundary. To reach Reserve, which is some 70 miles southeast of Datil, Barney would have to travel on U.S. Highway 60 through Magdalena across the top of the Plains, then turn southeast at Datil on New Mexico Highway 12 with the Plains visible for 35 miles or so. In short, Barney had a lot of traveling to do wherever he went. He traveled across, on, or around the Plains once or twice a week in good weather. I would estimate that he spent one-third to one-half of his time traveling, depending upon the distance he was away from Socorro. How widely did Barney travel in his work? All of Barney's known field work in 1947 was in or near Socorro, Magdalena, or in the high country, Plains area. He could probably reach any place in the Salado District in no more than two and one-half hours. Generally, the high country starts a few miles west of Socorro, as U.S. Highway 60 winds its way

*Figure 9.2. **Beth Danley** married Fleck in 1935. They had two children. The author met Beth in 2004 while speaking to a senior group in Magdalena. She told a friend "at first" Fleck and I were "curious who would travel all this way or call about this flying thing crash." She said over the years calls got very wearisome.*

up Sedillo Hill and through Socorro Canyon. The terrain flattens out south of Black Mountain. The highway skirts the north peak of the Magdalena Mountains and drops down slightly into Magdalena (elevation 6,548 feet). From there it runs straight as an arrow to several miles south of the Tres Montosas and angles onto the upper one-fourth of the Plains of San Augustin. U.S. 60 increases its elevation as it travels through the settlement of Datil and begins the ascent up White House Canyon towards Pie Town, Quemado, and the Arizona line.[11]

At Datil, New Mexico, Highway 12 heads in a west-southwest direction along the edge of the Plains. To the north of Highway 12 can be seen Sugarloaf Mountain, Wallace Mesa, Horse Mountain, and Horse Peak. A few miles south of the Mangas Mountains is the little hamlet of Horse Springs. About three miles from Horse Springs on a north-south axis lies the western edge of the Salado District boundary.

When Barney left Socorro and was out of the SCS office, Ruth usually recorded it. When he traveled west into his District and what we believe was north, south, or west of Magdalena, she sometimes noted the ranch's name or what ranch he was visiting or what town he was near that day. "The nearest town" in the west is merely a reference point. That is where the mail often arrives, but the ranch itself may be miles away. With limited space in her diary for each daily entry, it is believed if he had several places to go, she often generalized, "Barney to high

## UFO CRASH AT SAN AUGUSTIN

## BARNETT'S ASSIGNED DISTRICT 1946-1959

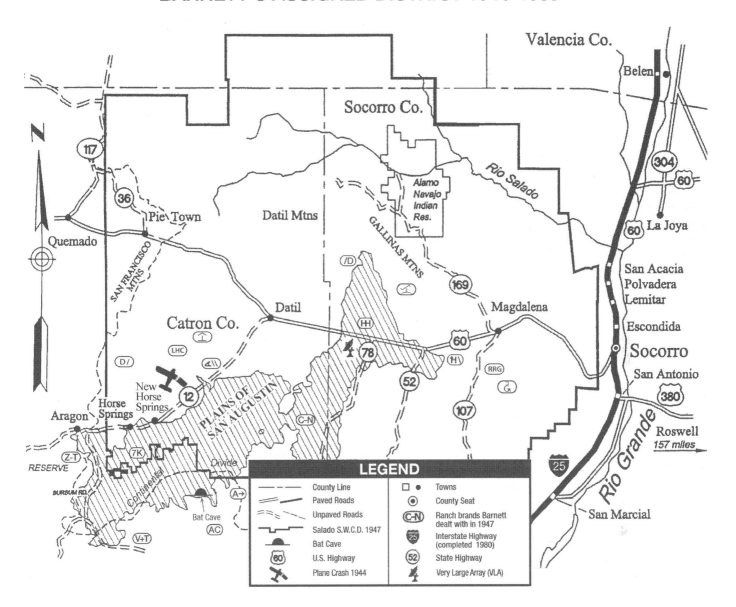

*Figure.9.3.* **Map of Salado Soil and Water Conservation District** *boundaries, showing relationship to the Plains of San Augustin, plus the boundaries of Catron and Socorro counties, as well as roads, towns and other landmarks. Barney Barnett was assigned to this large area 72 miles east to west by 70 miles north to south, working in this district from 1946 to 1959. Specific ranches Barnett visited (or may have visited) in 1947, marked above. Bat Cave, where the archaeologists made camp in early July 1947 to left of Legend box.*

country today." He made 42 trips to the high country from March to December of 1947. "High country" was the designation in those years for land on or near the Plains. Socorro was 4,600 foot in elevation and the Plains were 2,200 feet higher at about 6,800 feet. It is believed when Barney was in the field and was going to be late, he brought his government vehicle home. In 1947, this vehicle was a USDA owned 1938 G6 Ford pickup truck. When he went to the high country, he was usually home by 6 or 7 p.m. and sometimes as late as 9 p.m. On one occasion Barney spent the night with a rancher and did not get home until 4 p.m. the next day. Known ranches Barney traveled to and the number of trips in 1947, appear on the map in Figure 9.3.[12]

Ruth used one other term about Barney's whereabouts; that was "in the field." There were 67 of these entries made in Ruth's diary. We assume that in the field was no farther away than Magdalena (about a one hour drive from Socorro). This would allow him to go by the Magdalena office, check his mail and get some business done, and perhaps see someone. Sometimes he was late, but more often than not, he was done and back in Socorro by 5 p.m. Ruth would pick him up at the office if she had the car or he would drive home if he had it. Twelve of these "in the field" days were combinations of office and field work. Barney could arrange his schedule somewhat; for example, he almost always was able to come home Monday evening, clean up and get to his Rotary Club dinner meeting in the Presbyterian Church basement by 6:30 p.m.[13]

Barney visited most ranches about twice. Once in a while on a big project he went five or six times. The nature of his work involved some surveying, taking shots, driving stakes, etc., and going back later to inspect the work in progress or finished. Since it took 60 to 90 days for SCS paperwork to clear all the hurdles, some of the projects he worked on in 1947 were probably initiated in 1946. In a case like field leveling, the work may have been laid out by someone else. If it involved ground terrain slope, he as the project engineer may have had the final sign-off on it. In 1947 Barney and Ruth also took some eight trips to Albuquerque about 85 miles to the north. Barney had some occasional business there, but mostly they went to buy building materials for their new home. On one trip Barney had fish for lunch and got food poisoning. Ruth did not drive at night, but she had to on this occasion. They apparently got home safely.

As old Highway 60 crossed the upper part of the Plains, New Mexico Highway 78 intersected it near a place known as Augustin. A Civilian Conservation Camp (CCC) was located near here. About 100 men in the camp did public works projects. They built schools, set telephone poles, worked on roads and carried out numerous other projects in the area. The Works Progress Administration (WPA) program worked on some of the area arroyos to slow down erosion. One of the projects in connection with check dams was lateral arroyo channels. At high water these would fill and act as settling ponds when the water went down. World War II and the subsequent closing of the CCC program helped the ranchers see what good conservation work and practices could do. The SCS, through the individual districts like the Salado and Socorro Districts, carried on much of the conservation work begun by the CCC. [14]

At the war's end in 1945, the Soil and Water Conservation Districts faced a dilemma. They were very short of equipment for various conservation needs. To carry out many of the agreements with the ranchers, equipment was needed, including small bulldozers, mold board plows (to cut range furrows), seed drills, heavy tractors to pull harrows for field leveling, and road graders. It is believed that by 1947, with the availability of World War II government surpluses, the equipment shortages had eased up. Most of the work Barney laid out required some machinery and equipment. Alice Knight and her husband, J.B., owned an airplane. They visited Socorro three times in 1947.[15] The first trip into Socorro in 1947 was April 14. J.B. and Alice were still newlyweds and enjoyed the New Mexico School of Mines Senior Prom. There was a quick visit on October 10 and a Thanksgiving visit in November. Alice said they gave Barnett's friends rides and sometimes flew over the Magdalena-Datil area, and Barney would point out dams he had helped build. They also drove to some of the sites closer to Socorro.

## JULY 1947, THE DISCOVERY

Monday, June 30th was the beginning of the Fourth of July week. Ruth reported the day was warm with a sandstorm late in the evening. Barney went into the field in the morning and worked in the office that afternoon. He attended his usual Rotary meeting that evening. Tuesday, July 1st saw another warm day with Barney in the field and Ruth cleaning house all day, which most homemakers did after dust storms. Lopez, the man helping with the new house, prepared walls for plaster, and friends came over in the evening. Barney and Ruth retired early, and later it rained for an hour or more. Wednesday, July 2nd, "Barney went to high country near Datil," Ruth wrote in her diary.[16] This could have meant he went to one or more ranches between Magdalena and Datil, or he may have gone to one or more ranches southeast of Datil.

Barney was to later tell friends of his experience in early July of 1947. The July 2nd date seems to be the most logical because it places Barney definitely on the Plains, probably 25 miles or less from Datil. For July 2, 1947, Ruth wrote in her diary, "Cool and sunny." If the weather held for Barney, he had clear views out to the Plains from north to south, as he traveled towards Datil on Highway 60. When he went south of Datil (which we believe he did that day), he would have had a clear view of the Plains on elevated Highway 12 for 35 miles or so.

His niece, Alice, said she thought he was on a road near the Plains of San Augustin, "when he saw glinting metal off of some object in the distance." This would be quite possible, especially if it was morning between 9:00 a.m. and 11:00 a.m., when the sun was at the right angle for one traveling south. Barney said at first he thought the glinting metal he saw might be a plane crash and decided to go over to investigate. Alice Knight said she believed Barney had driven up to the site. A seldom-used, private back road ran close to the crash site. We are reasonably certain he would have used it.

Barney was a fairly astute man. He knew about the early rocket experiments at White Sands and he had some rudimentary knowledge gained from local college lectures on atomic energy and the atomic bomb. He knew enough from a military standpoint not to drive up to the "glinting object" directly. Since the object was in the arroyo, which was about 10 to 12 feet below the surrounding terrain, he probably planned his route to keep himself and his vehicle out of sight until he was very close to the object. He told Vern Maltais once that he had "approached it as though it was explosive, as anyone with military experience should have."

Vern Jean Maltais came up to Stanton Friedman after a UFO lecture at Minnesota's Beniji State College in October of 1978. They asked if he had heard of a saucer crash on the Plains of San Augustin seen by a witness and a good friend of theirs, Barney Barnett. Stanton said, "No, he hadn't." This is where the Barnett story began.

When Barney arrived at the site, he told Vern and Jean Maltais that it was not a plane[17] or a missile or a bomb, but rather some sort of metallic disc-shaped object some 25 to 30 feet in diameter. Barney described the outside of the craft as sort of a dirty stainless steel color. At some point, Barney told the military that he had touched the craft's surface. In a February 22, 1999 letter to the author, Maltais said, "In 1950 on our visit to Socorro, New Mexico and the visit with the Barnetts, I was told about the crashed saucer. He told me this craft had split open and four small beings were dead, some on the ground and some inside. Maltais continued, "Soon after he arrived at the scene, a group of archaeologists came up to look.[18] Then in no time at all, the military arrived and told them all to leave and that they had seen nothing." In a previous interview, Maltais had indicated that Barney had said the archaeologists had come up from the opposite direction. We know Barney had seen the "glinting metal" of the crashed UFO from New Mexico Highway 12. The opposite direction would have been to the southeast where Bat Cave is located.

As the crashed saucer story was being told to Friedman in October of 1978, Jean Maltais inserted, she "thought the archaeologists were from the University of Pennsylvania." This led researchers on a fruitless 21-year search, costing much valuable time and energy. In 2000, the author found the only team of archaeologists working on the Plains in early July of 1947 was from Harvard. This find was first announced at the Retrievals Conference at Henderson, Nevada in 2001.

Barney described the bodies to Vern and Jean Maltais as being small, with small eyes oddly spaced, no hair, and only four fingers. Their clothing was gray in color and one piece. No closures were visible, such as zippers, buttons, or belts. All the bodies appeared to be males. Barney told the Maltaises he was close enough to touch them. In a quote to the researchers in the Berlitz and Moore *Roswell Incident* book, Maltais quoted Barney regarding the bodies as saying, "They were all dead as far as I could see and there were bodies inside and outside the vehicle. "Descriptions of alien outerwear indicate that the surface was smooth and form-fitting. Maltais said they were dressed in tight-fitting, metallic suits as described in Chapter 13.19

Some gray fabric was found and sent to a well-known Chicago area forensic lab. They found the gray material to be rubber coated, and aluminum was listed as a major part of the fabric. This material was soft and apparently form-fitting. Was the aluminum in the cloth the "metallic suit" Barney told Maltais he saw.

Over the years Barney developed the theory that the bodies had been radioactive and he had contracted lung cancer from them. He relayed this to several people, including Harold Baca. Barney was a cigar smoker. The 1964 Surgeon General's first report on the dangers of smoking was not readily embraced by smokers of Barney Barnett's generation. Smoking may have contributed to his passing in 1969, and he did have lung problems thought to be related to a WWI gas attack.

In a videotape made for the Fund for UFO Research

*Figure. 9.4. Vern Maltais, 1920-2001, and his wife Jean, met Barney and Ruth Barnett during WWII, and they were good friends for many years. An Air Force master sergeant, he came to Mosquero, New Mexico in 1943 (where the Barnetts lived) with other military personnel to help with a wartime victory bond drive. While visiting the Barnetts at Socorro in February of 1950, Vern and his wife learned of Barney's Plains crashed flying saucer experience. Vern said the Barnetts never wrote about the crashed spaceship discovery. Barney thought it might jeopardize his government job. Photo courtesy of Vern Maltais.*

in 1993, Maltais recalled other details concerning what Barney shared with him. Included among his descriptions: "They were three and a half to four feet tall, very slim in stature and the heads were hairless, no eyebrows, no eyelashes. The hands were not covered and they had four fingers" each. This gives rise to an interesting question: If their hands were not covered, does this mean their feet were? If so, with what? The feet of "the little people" may have been covered with the one-piece gray body garment Barney described, similar to children's pajamas. The remnants of a small shoe found at the site may have been the extreme bottom of this garment or, if the shoes were worn over the garment, they were possibly blown off when the craft impacted.

When asked how Barney felt about what he had experienced, Maltais said, "He had no qualms about what it was. He said it (the craft) was a vehicle from outer space. There wasn't any question: the beings on there were nothing like human beings, similar, but not exactly."

In an interview that appeared in the 1980 *Roswell Incident*, by Berlitz and Moore, Jean Maltais was quoted as saying, "Barnett traveled all over New Mexico, but did most of his work in the area directly west of Socorro." (See map, Figure. 9.3.) This has led some researchers into thinking that Barnett had duties that entailed statewide travel, which may have put him at the Roswell crash site northwest of Roswell. I asked Vern Maltais about this. He said that he and Jean had talked about this over the years and she was referring to the Barnetts living in different places in New Mexico over a period of time. She told several interviewers this, but they never saw anything in print clarifying the previous words. The author's research has placed the Barnetts in four scattered New Mexico locations during their years there: Ute Park near Logan, Clayton, Mosquero, and finally in 1946, Socorro. Barney was also doing engineering work for the Soil Conservation Service in Springfield, Colorado, before moving to Mosquero in 1939.

Vern Maltais closed his February 22, 1999 letter to the author by saying, "Somewhere in my memory I think Barney had told me the disc was shipped out on some railhead, to where?" The nearest railhead to the Plains was at Magdalena. It was an Atchison, Topeka and Santa Fe spur that ran up from Socorro.

Summing up his friend Barney, Maltais said, "I would stake my life on Barnett's reliability, as I knew him well." This is quite a tribute to Barney from his friend Vern Maltais, who had served his country with honor during two wars.

Since I have become somewhat familiar with the Plains, some of the ranchers, the roads and something about travel patterns, I have a new appreciation of the vastness of the area and what could or could not be seen from what distance. I marvel at the oversimplified stories of how people found the Roswell and Plains crash sites. One could drive in these remote areas for days and not see a soul. I spent most of a day getting into and out of the Bat Cave area and never saw any other vehicle or even the dust from a distance. The notion that this or that group of people just happened onto the crashed UFO might happen if there was something that attracted them to the same spot, but for individuals or groups to accidentally merge together within a few minutes of each other would severely stretch the laws of chance.

While driving near the arroyo site one day, I was able to determine that a slight rise near where I was excavating was visible from a nearby road for over 700 feet. Barney talks about an object reflecting light a mile to a mile-and-a-half away. This fits the location of the site where I was working and found the artifacts. If Barney knew the way in via the back roads, it might have taken about 15 minutes to reach the site. There was one archaeological group that was known to have been working on the Plains a few miles away about that time. This was the Herbert W. Dick party. More about the archaeologists can be found in later chapters.

One thing is certain, the military arrived also. We do not know what base they may have come from, or from what direction, but they were probably aware of the flying saucer from radar and may have found the site from a plane search that directed the military vehicles and personnel to the site.

The military convoy may have been on the road for some time. It may have originated on the Alamogordo Air Force Base or the White Sands Proving Grounds or both. The mileage chart on the 1947 state of New Mexico Highway Department map shows Alamogordo; the town is nine miles from the Air Force Base. It was over 300 miles from the Plains. Much of this distance was on two-lane highways and some gravel roads. Over such roads, in those days, without New Mexico's present north-south Interstate 25, a military convoy would do well to make 35 miles per hour. It more than likely took at least six to eight hours to make the journey using military vehicles. I believe there were two or three vehicles, including a truck that arrived at the site.

Was the Plains discovery the first New Mexico UFO crash in 1947? It apparently happened before the announcement on July 8th of the Roswell event or at least was discovered sooner. As improbable as it may seem that such an event could happen, it must have been a mind-boggling experience for Barney. Over the years the military warnings to keep quiet have taken on a much more dramatic role than they may have had originally. Being asked to keep something quiet during wartime, apparently, was much easier than it might have been during peacetime. As events turned out, Barney had a more tongue-in-cheek attitude towards his silence. According to Vern Maltais, Barney told his boss, Fleck Danley, about the Plains discovery the same day. Danley, in later interviews, puts the crash event in early summer of 1947, which would fit the July 2nd scenario. Another military officer who corroborates Barnett's experience from a Washington D.C. perspective will be discussed later.

Reconstructing Barney's movements as he headed home on the day of the event would not be difficult. Thoughts and miles moved by rather quickly as Barney headed his pick-up towards Magdalena. He had completed his SCS agenda and taken the necessary stops, but after his morning's experience, it was not Soil Conservation projects that were on his mind. He could see Magdalena Peak up ahead. Over the pinon trees near the picnic table turnout, he could see the sunlit peaks of the Tres Montasos looming to the north. He had gone over and over these questions as he headed towards home on the afternoon of July 2. Would anybody believe him? Would the military find out if he told anyone? Could they have him fired or transferred? Would he have to sell his new house that he and Ruth had dreamed of? What would he tell her? Would she worry? If he did tell anybody besides Ruth, would people think he was crazy? Probably his biggest worry was his job. He had worked with the SCS since 1937 and had good job evaluations and an excellent reputation. For a man with a much better than average grasp of astronomy, where had these little people come from? All this and more must have gone through his mind.

Those who knew Barney said he was a fine engineer, well-organized with a calculating mind that weighed options. What were his options here? If job concerns were one of his first considerations, then he might head off any move by the military by reporting to his supervisor first. The military personnel were unusually gruff when they had warned the people at the site not to say anything. Were they bluffing or for real? He hadn't broken any laws; he was just performing his job as he drove past a spot he had driven by many times before. He was the SCS surveyor and was working with the landowner on a project, which would continue into the future. It would seem the first thing to do would be to talk to Fleck Danley.

The village, as the Magdalena City Council liked to call it, was no doubt having a typical quiet Wednesday afternoon. The western outskirts of town were marked by a few undistinguished houses and pole corrals here and there. Driving Highway 60 onto the main drag, things would seem normal. The Salome General Mercantile had its usual sidewalk displays and, across the street, the Phillips 66 station and Wesgate's Texaco-Firestone a block or so down were doing a little business, as usual. Barney may have seen a kid or dog or two in front of Harding's Drug Store. They had a good selection of penny candy and the village's only fountain with two scoops of ice cream for a nickel that few kids could afford. Down the street was the SCS office with a few dusty vehicles, including Fleck's truck, parked on the side. If it hadn't been there, Barney would probably have gone on into Socorro. Now was as good a time as any to tell Fleck about the crashed saucer he had discovered. Even though he wasn't sure how Fleck would take it, he needed to tell someone.

In the Berlitz and Moore book, Fleck told researchers that, as near as he could recall, Barney came into the office one afternoon ". . . all kind of excited," and exclaimed, "You know those flying saucer things they've been talking about, Fleck. . .?

Well, they're real." Barney said he had just had a look at one of them.[20]

There are different versions of the event, as seen through the eyes of Barney, and from Danley, who was first interviewed about 1979. Both men's accounts tell us something of their temperaments. Barney told Vern and Jean Maltais that when he mentioned seeing a crashed flying saucer to Danley, that his boss said, "Forget it. I don't want to hear about it!" Danley's account is a little less charitable toward himself. He remembers saying "Bull _ _ _ _!" If Danley said this, he broke a western courtesy tradition here. Apparently, Danley thought better of his reaction and asked Barney about it several days later. This may have been after the Roswell story broke on July 8th. During the second conversation, Danley related to interviewers that Barney briefly described the craft and said he did not want to talk about it anymore. This incident may have had some bearing on to whom and when Barney spoke of the incident again. Years later Fleck recalled Barney as a good engineer. He required little or no supervision. The ranchers respected him, and he was instrumental in getting quite a few SCS projects started.

Although this research was begun before Fleck Danley died, I did not have the opportunity to interview him. I was in touch, however, with some who knew him, as well as his wife Beth who kindly sent his photo for this book. I learned that her husband Fleck had gone with Barney, within a month or so after the Plains crash, to the site. It is believed that they were there to determine land boundaries. Apparently, Barney and Fleck were both concerned about the land areas managed by the Soil Conservation District, and what effect it would have if the government confiscated Plains grazing land as they had done in the White Sands Proving Ground area to the east. In the author's investigations, there are indications that Fleck had gone later to Albuquerque concerning this matter. He apparently talked to administrators at the Soil Conservation Service (SCS) and the U.S. Department of Agriculture (USDA). Beth indicated to another researcher that she thought Fleck had gone to the crash site on other occasions in the late 1940s, but could recall very little other information.

According to Ruth's diary, on July 2nd,[21] while Barney was in the high country, Ruth attended a luncheon and spoke to a friend who wanted to rent their house when the Barnetts moved into their new one. Ruth played bridge and lists Barney coming in from Datil at 6 o'clock. After supper their friends, the Thompsons, came over for awhile. During the rest of the week, which was warm, Barney was in the office. They received welcome mail from friends, and the new house was being plastered. On Friday, the 4th of July, Barney worked on the house all day, although he was not feeling very well and Ruth reports that they spent a quiet day. Their friends, the Ames, came over for dinner and then drove out to a hill west of town to watch a fireworks display over the city. If Barney had his Plains experience that week, we can only wonder what he might have thought as he looked into the night sky on what very well could have been one of the most important weeks of his life.[22]

**THE ROSWELL EVENTS AND BARNEY'S TRAVEL**
On the Foster ranch about 50 miles north of Roswell, New Mexico, a ranch hand named William (Mac) Brazel found a large field of metallic debris and some other material sometime during the three-week period between June 14th and July 5th, 1947. Brazel brought some of the material to the Chavez County Sheriff's office on or about July 6th. Two men from the Roswell Air Force Base, Major Jesse Marcel and a counter-intelligence officer named Sheridan Cavitt, accompanied Brazel back to the ranch to investigate. On or about July 7th, the three traveled to what is known as the debris field, where they gathered more samples of the unknown material. It is believed that the crew of Air Force men combed the site and picked up all of the remaining debris, which was eventually shipped to the Air Materials Command Center at Wright Patterson AFB, Dayton, Ohio. Later, Mac Brazel's adult son found some debris and collected it in a cigar box. He showed it off at Wades Bar in Corona, New Mexico several times, and the Air Force (OSI) Office of Special Investigation confiscated it. In the next few days some of the pieces from the crash site were flown to the office of Brigadier General Roger Ramey's 8th Air Force District at Fort Worth, Texas.

It is believed by many that, within a day or two, the Air Force had located the main part of a crashed flying saucer, complete with several alien bodies. The initial debris field find received international publicity when the base public relations officer, Lieutenant Walter Haut, released information to the press about the material Bill Brazel found.[23] With much maneuvering and intimidation of some witnesses, the Air Force was able to keep the crashed saucer event quiet and at the same time discredit the debris field story, saying the material came from a wrecked weather balloon.

As far as we know, Barney traveled only once east of Socorro and across the nearby Rio Grande River in 1947. That occasion was to Roy, New Mexico, from January 27th to January 31st for some SCS business. He rode to Roy with an associate and returned on a Trailways bus. He did occasionally go to nearby towns on what it was assumed was SCS business, but almost all of his work was done in his designated District as indicated on the map in (Figure 9.3). Ruth's diary entries were mostly done the next day for the previous day. Often she mentions what they did in the evening and on several occasions things that happened after they went to bed, such as the weather, mosquitoes, etc. She probably knew where he said he was going before he went to work and he no doubt talked about it when he arrived home. I believe she recapped the previous day's events in her diary before she started her daily routine.

Ruth's diary was actually a 1947 daily appointment book from the Mitchell Insurance Company of Dalhart, Texas. With two entries per page, each date's entry occupied only one-half of a page which was 4 $3/4$ inches wide by 7 $1/2$ inches long. One of the big questions that arises is: Why was Barney's flying saucer Plains discovery not reflected in Ruth's diary? There were several reasons for this. I am sure that Barney was greatly intimidated at the Plains crash site by the military presence, and there were indications that his boss Fleck Danley was very concerned also. I asked Vern Maltais if Ruth Barnett ever mentioned it in any of her letters to them. He said that no, she never had, as far as he could recall. It was understood that, since Barney had waited three years to tell them in person, he obviously did not want anything written about it. Vern said the incident was very troubling to Barney at the time and he was afraid that his mail might be monitored. Alice

Knight's, Ruth's niece, thought the stress of the whole episode may have contributed to Barney's bad heart attack in 1949.

Barney told the visiting Army Lieutenant William Leed in 1967 that he had been warned by high and low-level government people to "Shut up about the crash." The high level people may have included Air Force military and civilian personnel. But who would the low level people have been? Was it Fleck Danley? Where there others in this chain of supervision interested in keeping this quiet? We are sure Fleck (Barney's boss) made a query about the military intentions on the Plains to the higherups in Albuquerque or perhaps Denver.

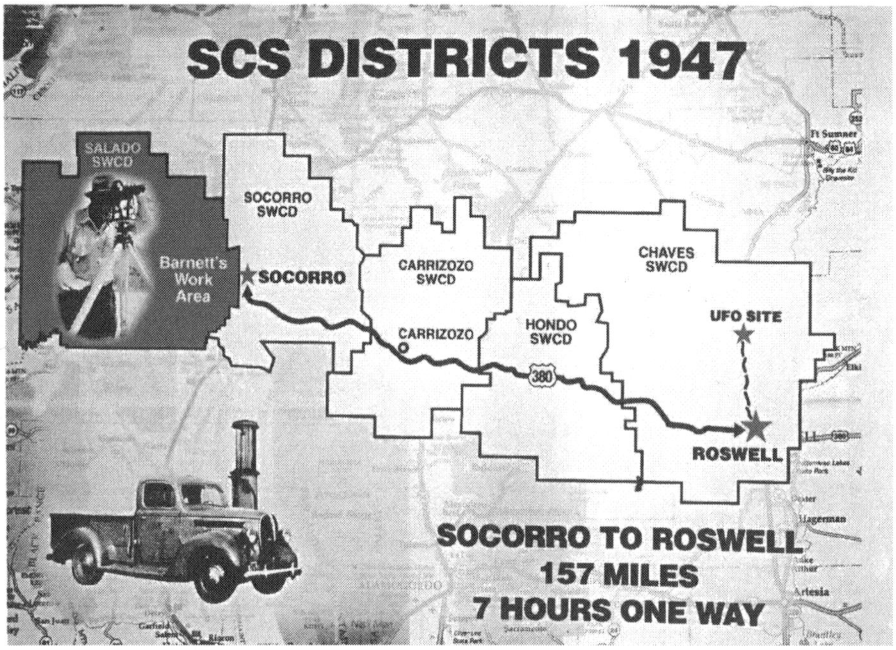

As mentioned before, the government was condemning and taking thousands of acres of land for the White Sands Missile Range project west of Socorro. We believe that Barney did the right thing by going to Fleck Danley after he had been hassled by the Army at the crash site. We also believe Fleck's subsequent contact with the USDA in Albuquerque may have initially been comforting, but eventually backfired, especially if they had to issue warnings to Barney to keep quiet. These concerns, no matter which direction they came from, may have been diffused when Barney and Fleck ran a survey on the crash site. They may have found out, as did the author, that the land was about 100 yards inside some state land boundaries, and the government could not have acquired it in any case if they had wanted to.

The question arises: Why did Barney tell others after he was asked by the military not to say anything? I, frankly, have no answer to this. We do not know whether he even agreed in principle not to say anything. Being a witness to something this big must have weighed heavily on him. In all cases, as far as we know, Barney told only one associate, Fleck Danley, and family at first (Ruth, Alice Knight, and J.B.), then a few good friends that he trusted over the years, including the Maltaises and several others.

A secret such as this must have been hard to keep, but as far as we know, Barney did not reveal to anyone except Fleck Danley, exactly where the crash occurred. There may have been one other that saw the crash site— that was Alice's husband, J.B. On several occasions after they learned about the crash, J.B. and Alice flew in to visit the Barnetts in Socorro. There are some family records that indicate Barney and J.B. flew up to the "high country." And they no doubt flew over or near the crash site. I am not sure Barney and Ruth could have anticipated the ramifications of this experience in light of what we know in subsequent years, although keeping things out of print would have been a prudent thing to do. Alice Knight said she never saw anything written about it, nor did Barney tell any other relatives about it. She and Ruth corresponded several times a month.

Barney had started with the Salado SWCD in 1946 and had an excellent reputation. Stories of an unbelievable alien spaceship could hamper him in his dealings with peers and the ranchers with whom he worked. After fifty years or so, ranchers today might be more open to aliens and spaceships, but certainly not in 1947. One lesson civil servants learn sooner or later is not to make waves or call negative attention to themselves. Barney expressed at least one of these fears to Vern Maltais. Another reason Barney's experience may not appear in Ruth's diary is that apparently Barney and Ruth made a decision not to ever write anything about the event. Also, of course, the diary was primarily a record of Ruth's activities as well as Barney's departure to work and his arrival home each day. Her diary seldom mentioned anything about his work, only the area and occasionally a specific ranch where he was located on a given day.

*Figure 9.5. **Socorro to Roswell 1947?** There has been some effort by some UFO researchers in the past to connect Barney Barnett with the Roswell UFO crash event in the first week of July 1947. In those days, U.S. Highway 380 was a winding, twisting, partially unpaved, seven-hour, one-way drive from Socorro. Once he reached Roswell he would still have to drive 70 miles north and across country to the debris field of the Roswell crash site, a one-way trip that would take nine hours and was some 225 miles out of his district.*

Some of the Roswell writers found it expedient to bring Barney over to the Roswell area with some archaeology students to discover the craft and bodies found there. This was when the Roswell writers were trying Barney out in different scenarios. Barney's experience was compelling and with the addition of the bodies, made a rather complete story. The initial description of the bodies is still the basis for describing gray aliens today, although the eyes became much bigger. After UFO researchers had exhausted what they thought were all possibilities of locating the archaeologists Barney had seen at the site, they began to move Barney around, speculating his involvement in various theories.

Another possible theory involves Barney on July 8. Ruth's diary for that date had Barney going to Pie Town about 75 miles from Socorro and returning late. The theory being tried out here was that Barney had gone to the Roswell area, discovered the crash and bodies, tried to help the military cover up the crash and had come home and lied to Ruth about traveling west instead of east. The problem with some of these bizarre scenarios is that Ruth always knew where Barney was going so they could communicate in case of emergency.

As alluded to before, Barney, when working in the Salado District, usually visited a high country site two times. The first visit may have been to lay out the work with surveying equipment or otherwise give recommendations to the landowner on how best to accomplish his goals. There was usually a second visit to the same site. In the case of Pie Town trips, he may have had several projects to lay out with different landowners and had to check progress on his next trip. Barney went to Pie Town on Tuesday, May 1st and returned home late. He visited Pie Town again on July 8th, which is consistent with his two trip pattern. On July 8th, he also got in late, at 8:30 p.m., according to Ruth. The duration of his trip to Pie Town is also consistent with the first trip. To reinforce Ruth's entry concerning Barney's July 8th trip to Pie Town, she mentions that he went to the high country again on July 9th. Again, this would mean he went back to the high country for the second day in a row.

The route to Socorro from Pie Town is on an east-west main highway, U.S. 60. This main road in the area cuts across the top of the Plains between Magdalena an Datil and goes to Pie Town and Quemado to the Arizona line. There is an interesting sequel to Pie Town. When the economy picked up in the mid-1990s, one or two new businesses bloomed in Pie Town. Among the businesses was a nice little restaurant serving very good pie. I was talking to a waitress one day who was the wife of the owner. When I showed her some of this research, she exclaimed, "So, you are the one talking up our business."

In a 1991 book, UFO researchers posed this question: Had military officers at the Roswell crash site talked with Barney Barnett there and asked him to cover his tracks by saying he was in Pie Town that day (300 miles to the west)? I find it somewhat a stretch of the imagination to believe he would be involved in such a ruse and would ask Ruth to falsify her diary entry, thus providing a future alibi for himself in case the Roswell event should surface. Would Barney say goodbye on the morning of the 8th and say that he would be at such and such a ranch in Pie Town, then actually head 175 miles out of his district to the Roswell area – a seven or eight hour drive one way? Why should he? Another theory that attempts to work Barney into the Roswell scenario was put forth, indicating that Barney discovered the Roswell crash and was asked to move his discovery over to the Plains.

Figure 9.6. **William Leed,** upon his graduation from St. Lawrence University and their ROTC program, in 1962. Five years later, First Lieutenant Leed drove to Socorro, NM to talk with Barney Barnett about his UFO crash experience. Bill Leed learned of the Plains crash and obtained Barney's name from Pentagon sources. He served in the signal corps in Korea and with the third infantry division at Ft. Hood, Texas. Yearbook photo courtesy of reference library, St. Lawrence University.

Barney's discovery of the Plains crash was five days before the Roswell crash was initially reported. If Barney had anything to do with the Roswell crash, it would mean that he had falsified the story to his boss the day he found the Plains crash with bodies. He would have also had to lie to his wife when he got home that evening, then to his niece and her husband at Thanksgiving 1947.[24] He also shared the story with friends Vern and Jean Maltais at the time of their visit in 1950.[25] He then repeated that story to several friends, including Harold Baca, his neighbor, and to Col. Bill Leed 22 years later. It is obvious here that Barney and Ruth did not want to put the story in writing. This coincides with statements by Vern Maltais and Alice Knight that they had never seen anything in writing either. Barney's story to those few he shared it with was consistent, sincere, and lifelong.

**CONFIRMATION**

Although on active duty in the Army, on this trip Bill Leed was dressed in civilian attire. It had been a long, weary journey from Fort Hood, Texas near Killeen. Bill, at that time was a first lieutenant and assigned to the U.S. Army's 3rd Corps, 2nd Infantry Division.[26] The trip from Fort Hood through Texas, and on west to Roswell, had been uneventful. Bill had built up some leave time and was taking a vacation. It was hot in the Southwest at that time of year, and Bill recalls that his car had no air conditioning. But the wind wing of the car had at least supplied moving warm air while he drove.

It was August of 1967, and Bill had just completed a hitch in Korea where he had served at Camp Howse, north of Seoul. He was in his fifth year of what would be twenty-eight years of service to his country. In 1962, he had graduated from St. Lawrence University in Canton, New York, with a major in history and government. That same year he had also finished the ROTC program and had been commissioned a second lieutenant in the Signal Corps in the U.S. Army Reserve.

Camp Howse, Korea, seemed a world away from Socorro, New Mexico, and it was. Bill was now on his way through Roswell on U.S. Highway 380 to Socorro. [27] As we related earlier, in the early '60s, Bill had been stationed in Washington, D.C. There, he had a brief conversation with a higher-ranking officer from the Pentagon, when the subject of UFOs had come up. The officer had said, "Yes. We know all about that." Leed related how, with his rank of second lieutenant, he was "low man on the totem pole," and he really did not want to press the matter. The officer continued, saying, "If you really want to meet a man who touched one (a UFO), go see Barney Barnett in Socorro, New Mexico." [28]

This tells me that Barney's name, address, and other particulars, and probably relevant material from other Plains crash witnesses, went through channels and were on file in Washington, D.C. and the Pentagon. There is no way the information that Bill Leed received, in the early 1960s concerning the 1947 event, could have come to him other than through higher military channels. It is interesting to note that Bill Leed's route from Ft. Hood, in the Texas Panhandle, where he was stationed at the time, to Socorro, took him through Roswell.

Although there were alleged witnesses to the Roswell event who were living there in 1967 (who would not come forward for another 12 years) it was not suggested that he talk with anyone there. Only Barney Barnett's name was given as a UFO crash witness to an event 160 miles due west. Knowledge of this also tells me that for security reasons, crash sites were compartmentalized on a "need to know" basis, accented by military base rivalries. We know that in 1947 there was little cooperation between Army Air Corp, (which would shortly become the USAF) and other branches of the service. The Pentagon officer who Leed talked to may have had official reasons to know about the Plains crash, but was not privy to the Roswell crash story. That would not surface for another 10 years in the Jesse Marcel interviews of 1978, leading to the *Roswell Incident* book.

Coming through the pass on Highway 380 between the Capitan and Sierra Blanca Mountains was a cool relief from the flat plains country throughout west Texas and eastern New Mexico. Highway 380 terminated at San Antonio at the junction of the main north-south highway through New Mexico in those days, U.S. 26. Turning north, Bill would be only 13 miles from Socorro. Before he left, Bill had called information for Barney's address at 520 Park Street. Entering Socorro in 1967, one beheld a different scene than today's modern interchange off Interstate 25, built in 1980. Bill would have ended up on Socorro's main drag, known as 85 (Highway 85) or California Street to the locals.

Bill felt a bit nervous at the reception he would receive, and did not know if the Barnetts were even home! So he decided to just drive by the house. The Park Street house was not difficult to find. The neat white stucco house stood out in an otherwise average block of homes and a vacant lot or two. A railroad spur crossed Park St. just below the house. Across the street was the not-too-attractive site of the Standard Oil bulk plant. Bill pulled up to the house and, for some reason, rolled up all his windows.

The house was situated lower than the sidewalk, so he went down several steps to reach the door. He reported that the door was opened by a thin man who was quite surprised at the visit from the stranger, but courteously "invited me in." [29]

When interviewed in 2003, Bill rather apologetically admitted, "I was a complete interloper. I just came in out of the blue!" He said he just wanted to "satisfy my curiosity." After some initial pleasantries, "Barney asked to see my identification," reports Leed. Bill explained that he was just an interested party, but showed him his military ID.

Leed began the conversation by asking if it was true about Barney's finding a saucer crash. Barney replied that he had come upon this crash site and that it was genuine. Bill says, "I took him at his word, and for me, that cleared up a lot. I knew the military knew something about it." Bill said, "It didn't occur to me to ask about any occupants of the craft," which today we call aliens. "If I'd have had any idea they were involved, I would have asked him about them." Bill said he didn't find out about the aliens Barney saw until the first Roswell book came out in the early eighties. Leed stressed that he was there less than 15 minutes total, and the talk of UFOs was limited to four to seven minutes. As the conversation progressed, Leed said Barney seemed a little anxious. "After 15 minutes or so, I excused myself. I think at first he was afraid that I was one of those types who came by to keep him silent . . . I wasn't! That's probably why he was so surprised to see me, and let me in." [30]

Bill indicates that, in retrospect, Barnett was relieved that he (Leed) was not a military official. Apparently, the fact that Bill had encouraged him to talk when he had been warned not to, made Bill apprehensive also. Bill remembers returning to the car with the windows rolled up, and getting in it was like sitting in an oven. Here was First Lieutenant Bill Leed (who hoped to make captain some day) soliciting information about one of the nation's top secrets he was not privy to.

In his working years, Barney had been of average build, but relatively short. He had played halfback on his college football team before World War I. The "rather thin man" Bill Leed saw that day was Landon Grady (Barney) Barnett now in declining health. Throughout his life, Barney often showed interest in things scientific and was still pursuing his lifelong interest in astronomy. He was also quite interested in the Apollo missions in the late sixties. About that time, Barney's health began to go downhill with compounding respiratory problems. Then, sometime after Christmas 1968, his condition took a severe turn for the worse. His photographs in the last five years of his life reveal that he was losing a lot of weight. In March of 1969 he fractured some ribs, probably in a fall. It is believed that the rib problem affected his breathing, and he then contracted an acute inflammation of the lungs with resulting high rates of pulse and respiration, coughing, and severe chest pain. With the broken ribs, coughing must have been extremely painful.

Several people also recalled that Barney suffered from throat cancer as well. The reader may recall a conversation between Barnett and Harold Baca the previous summer. The throat condition affected his voice, and Baca said he spoke very softly about "the little people" he had bent over on the Plains in 1947. Although medical records indicate the cause of death as bronchopneumonia, Barney also had arteriosclerosis, which may have been linked to a heart attack in 1949. The *Socorro Chieftain* of April 22, 1969 reported that Grady Landon Barnett, 76, died on April 11 at Socorro General Hospital after a lengthy illness." Ruth was at his side when he slipped away that evening at 8:00 p.m. They had been married for 52 years. Three months later, the Apollo 11 crew landed on the moon. Alice Knight said with Barney's interest in astronomy, he followed the NASA Apollo missions closely. He would have been thrilled at the news of the first moonwalk.

Some 10 years later in 1979, Bill Leed heard Stanton Friedman and Bill Moore talking about the Roswell crash on a Toronto radio station and contacted them. Friedman said in 2003 of the broadcast "that Barnett's name hadn't come up in the radio program. How did the colonel (who talked to Leed) know details about somebody who had such a low profile (Barnett) unless he heard them from official channels?" Col. Leed is an excellent example of an independent corroborating witness. Testimony from this type of witness has greatly influenced the outcome of important legal matters. Leed retired from the U.S. Army Reserve in 1990 and worked in his family's steel business and was also in banking for a time. In 2008, Bill Leed was living quietly in upper New York State, enjoying his retirement and traveling to Florida in the winter.

## ANOTHER CONFIRMATION?

Another confirmation may have come from a most unexpected source. In 1979, a journalist who learned about Friedman's initial contact with Marcel, interviewed Maj. Jesse Marcel, Sr., the ex-intelligence officer from the Roswell Base. Marcel was the first military man to see the Roswell UFO wreckage. A written transcript of the interview was found on the Internet. [31] There was an astounding revelation about a second crash site, which we believe alludes to the Plains of San Augustin crash. It was just a few words, but we believe they were very significant. Marcel was talking about the debris he found on the Foster Ranch and said on page 7 of a 10-page interview:

A: Lord, yes, about as far as you could see. Three-quarters of a mile long and 200 to 300 feet wide. I tell you what I surmised. One thing I did notice: nothing actually hit the ground, bounced to the ground. It was something that must have exploded above the ground and fell. **And I learned later that, farther west, towards Carrizozo, they found something like that too. That I don't know anything about . . . the same period of time — 60 to 80 miles west of there.**

**Q: Ranchers found something similar out there?**
**A: I think it was discovered by some surveyor out there.**

Some questions arise. Would Friedman, who set up the interview between Pratt and Marcel at Houma, Louisiana on December 8, 1979 have told either man about the information he received from Vern and Jean Maltais, concerning Barnett and the Plains of San Augustin UFO crash? Although the Roswell investigators had knowledge of Barnett's report, Friedman indicated that he, Berliner, and Moore were very busy getting the final phases of the *Roswell Incident* book to the publisher and that details were fuzzy. This information about a possible second crash did not appear in the article written by interviewer Bob Pratt which was in the *National Enquirer* in February of 1980. Friedman has no recollection of the material being discussed, but is uncertain about it.

Marcel's statement is sequential, as if he is recalling the incident from decades past. If the material had been recently given to Marcel, he would have certainly been more specific about where the craft crashed (other than 60 to 80 miles away), the bodies found, and would probably have mentioned how he learned it. People in New Mexico were very much aware of the Plains of San Augustin in those days. They are a major New Mexico landmark and the famous VLA (Very Large Array) radio-telescopes were being constructed there at that time.

Marcel Sr. uses a collective "they" when he said, "they found something like that too." Might Marcel's reference to "they" have been the archaeologists in Dick's party and/or would they have included the military when they drove up? Mentioning "they and a surveyor" indicates Marcel thought others were at the crash site also. The Mountain States phone book for the Socorro and Magdalena areas in 1947 lists no surveyors. It was generally known that soil conservation engineers like Barnett did occasional private work, and he was frequently in the Plains area.

Marcel says in his interview that the crash was in "the same period of time." Assuming this was the Roswell crash, UFO researchers had not established a possible Plains crash time line until Barney's boss, Fleck Danley, recalled "early summer 1947," which was strengthened by Ruth Barnett's diary of her husband's work and travel found in 1990. It was more than a coincidence that Marcel cited the same time period 13 years before other civilians came forward as witnesses. Only with military information could Marcel, Sr., have referred to the same time period in 1979.

When Marcel was referring to metallic crash debris strewn over a wide area and said "They found something like that" might he have been referring also to the metal pieces and foil shards we found at the arroyo crash site with the "skewed isotopes" and the strange, exotic coatings?

Major Marcel, who retired from the Air Force as Lt. Col. in 1950, was operating a TV repair shop in Houma, LA in 1979. Lets consider the following points:

1. Marcel's word "surveyor" is quite telling. The first use of the word "surveyor" came from the author's initial writing on this subject in 2001, or so. We knew that Barnett was a surveyor from the onset of this research. The extent of Barnett's expertise came from his 1941 civil service application. His surveying duties, as previously mentioned, are again listed on pages 64 and 65. He was not only an excellent surveyor but supervised survey crews for most of his early engineering work in Texas and Oklahoma after World War I. After a diligent search, we could find no UFO research papers, articles or books, between 1980 and 2005 that allude to Barnett as a surveyor. When referred to, he was always called a civil engineer or soil conservation engineer.

2. We know through Barnett and others that the military thoroughly questioned the civilians found at the crash site. In this intensive questioning, Barnett would have certainly let it be known that he worked for the SWCD and was a surveyor. He could prove this. if need be, by the surveying equipment he kept in his truck. If none of the *Roswell Incident* writers and researchers knew Barnett was a surveyor, and the term was not used for 25 years or more in any literature, then Barney's title as surveyor must have come through military channels to Marcel — not through *The Roswell Incident*

writers Berlitz, Moore, and Friedman, who had no direct knowledge of it.

3  Information coming from Linda Marcel (Jesse Senior's daughter-in-law) indicates he mentioned another "crash out west" several times before 1978 and once shortly before he died.

Is the Plains of San Augustin that site? Taking into consideration all of the exotic materials found there and the scientific reports, we definitely have a UFO crash site. It is quite possibly the site Marcel referred to. There is additional irrefutable evidence in Barnett's favor. One of the clinchers for me that Barney Barnett was present at the Plains crash site was his description of the aliens found there. He began describing them in the summer of 1947 to his wife Ruth, his niece, Alice and her new husband, J.B., at Thanksgiving time that year and in 1950, to Vern and Jean Maltais. The Maltais' descriptions were given twenty-nine years later (in 1979) to the authors and researchers of the *Roswell Incident*. Much of what we know today about the aliens comes from Barnett's 1947 description. More material came out from credible sources in later years, including Col. Corso's book and the SOM1-01 operations manual printed in various UFO literature. Barnett's alien descriptions were amazingly accurate, considering that he had less than a half-hour at the site before the military arrived. There was, however, a 32-year interval between his experience and the Maltais' reporting it to Stanton Friedman at Bemidiji, Minnesota.

The aliens Barney saw on the Plains were covered with a one-piece gray suit, with no zippers, belts, or buttons. In addition to the gray suit of the aliens described today, he got five other details correct: a small body, large head, small eyes oddly spaced, no hair, and four fingers. One of these details was repeated to his neighbor, Harold Baca, twenty years later in 1967 when Barney spoke of bending over "the little people" on the Plains of San Augustin. There were simply no descriptions in the media in 1947 or before, of little aliens of any size, description, or color. How could Barney (the first civilian we know of) be able to describe the aliens so accurately unless he had seen them?

Barney's description of the alien eyes as "small eyes, oddly spaced" is at odds with today's alien description of them having large, black, wrap-around eyes. If today's aliens, mostly reported in abduction cases, have the large, wrap-around eyes, then are we looking at EBE clones or laboratory-created beings? Two questions then must follow:

1. Assuming that the aliens then and now have similar origins, were the 1947 arroyo crash aliens of natural origin? and

2. Are the aliens with large black eyes part or wholly artificial clones, or some combination?

## CHAPTER 9
### REFERENCES

1  *The Socorro Chieftain*; Thursday, August 14, 1947; Socorro County Publishing Co.; Hugh Thompson, publisher; p. 4.
2  Ibid.
3  *Chieftain*; September 25, 1941; op cit.
4  *Socorro-Magdelena Telephone Directory*; The Mountain States Telephone and Telegraph Co.; Telecommunications History Group, Denver, Co.
5  Farr, Dave; Art Campbell private correspondence, 1998.
6  *The Socorro Chieftain*; op cit.
7  *The Socorro Chieftain*; Thursday, January 29, 1948.
8  Ibid.
9  *The Socorro Chieftain*; op cit.
10  Barnett, Ruth; Diary for 1947.
11  Highway map of State of New Mexico; 1947.
12  1947 Salado SWCD Boundaries Map; The PSA and other landmarks; p. 86.
13  Op cit.
14  Campbell, Art. Correspondence with USDA and NRCS (The former SSWCS).
15  Campbell, Art; Private correspondence with Alice Knight; 1996-2002.
16  Diary entry; July 2, 1947.
17  Campbell, Art; Private correspondence with Vern Maltais.
18  Notarized affidavit of Vern Maltais; Bemidji, MN; No. 7; April 23, 1991.
19  Ibid.
20  Berlitz, Charles and More, William L.; *The Roswell Incident*; Grosset and Dunlop, New York, NY; First printing 1980; p. 61.
21  Diary, July 2.
22  Ibid.
23  Randle, Kevin D. and Schmitt, Donald R.; *UFO Crash at Roswell*; Avon Books, The Avenue of the Americas, New York, NY; 1991; p. 90.
24  Diary; Alice Knight and JB visit; November 26, 1947 to December 1, 1947.
25  Campbell, Art; Personal correspondence with Vern Maltais; 1997 - 2001.
26  Campbell, Art; Personal correspondence and contact with Col. Bill Leed; 1998-2009.
27  Road Atlas; AAA Travel Publishing; New Mexico, p. 84; Texas, p. 11-12; 1994.
28  Campbell, Leed correspondence, loc. cit.
29  Ibid.
30  Ibid.
31  http://wiki.razing.net/ufologie.net/rw/w/jessemarcel.htm

These reference were consulted during this chapter's research

*Brand Book of the State of New Mexico Registered for Cattle, Mules and Asses.*

Certificate of Death, Grady Landon Barnett; State of New Mexico, Socorro, NM. April 11, 1969.

*MUFON Journal*; Stanton Friedman; November 2003; No. 427; p. 18-19.

*Lincoln County News and Carrizozo Outlook*; Era B. Smith, Editor/Publisher; Carrizozo, NM; July 11, 1947.

# CHAPTER 10

## LETTER FROM HARVARD, THE ARCHAEOLOGISTS & THE METEOR

### WRITTEN OFF

The evidence was clear "the best evidence indicates that Herbert Dick and his group were not at Bat Cave in early July 1947. No other archaeologists have been demonstrated to have been on the Plains in early July 1947." [1]

In 1991 before Dick died, Stanton Friedman wrote: "The Bat Cave report noted that Dick had done a preliminary examination for three weeks in the summer of 1947 with his wife, her brother and a friend, and a couple named Brown." Only Dick, his wife, and her brother were available when the interviews took place. Friedman continued, "Their testimony indicates that they were not at Bat Cave in 1947 until at least mid-July." [2] Thus concluded the report on *The Plains of San Augustin Controversy*. This report summarized the war of words that divided the UFO researchers in the early '90s. Of course we have more information now, some of which will follow.

Dick's involvement in the controversy is a keystone that supplied answers for our later research. On one side of the controversy over a crash on the Plains of San Augustin was nuclear physicist Stanton Friedman and author Don Berliner, who co-authored *The Roswell Incident*. On the other side of the conflict, maintaining that there was no Plains crash, were Kevin Randall, Don Schmitt, and their researcher at the time, Tom Carey. Jeff Morris, who was Dick's brother-in-law, told Carey and others that they were not at Bat Cave in early July, but were later. Could Dick and his party have been on their archaeological survey across the Plains earlier in July? Dick's team was emphatic that they were not at the Bat Cave site until mid-July. On June 23, 1990, Kevin Randall interviewed Dick. He said, "If I knew anything, I would tell you." Growing impatient with all the calls, Dick apparently said this to several researchers since 1989 up until the day he died in 1992.

It seemed to researchers in those groping years of UFO studies that Dick's presence or knowledge of the Plains UFO crash would also enlighten researchers as to Barney Barnett's involvement. If the UFO researchers in those days thought that Dick's academic reputation and standing at a major university would lead to some kind of disclosure concerning the crash and/or seeing Barnett at the crash site, then they would be very disappointed. Dick was very skillful at maneuvering the interviewers away from the dates he and his party were actually at Bat Cave, and would subtly point researchers in other directions.

Dick had been with Army Air Corp intelligence in WWII and had served in England. Kevin Randle, one of the main UFO researchers was also a commissioned officer with an intelligence background in the Air Force. Although one generation apart, it is interesting to note that two intelligence officers were taking the same side of a key argument for different reasons. Dick said he was not on the Plains in early July 1947 and convinced Randle and others of it. Randle maintained (and still does at this writing) that there was no UFO crash on the Plains of San Augustin.

*Figure 10.1. **Dr. Herbert Dick,** 1920-1992, was a well-known dedicated anthropologist and college professor. He took undergraduate work at the University of New Mexico and obtained his doctorate from Harvard in 1955. He was best known in his field as the discoverer, at Bat Cave, of maize dating back more than 4,000 years. It is more than likely that, during his 1947 stratigraphic survey, members of his party were seen by Barney Barnett at the Plains crash site. Photo courtesy Adams State College.*

One was less than truthful, the other wrong. This basic oversight affected UFO research for many years.

In August of the millennium year 2000, we received one of the most important pieces of material in this book. I had made a contact to the Harvard University about six months earlier to request faxes of any material that might be relevant to my investigations. On this particular day, I received a call from the business that had agreed to receive any incoming faxes for me. The voice on the other end of the line informed me that they were cleaning out their "dead" file and had located a fax for me to look at before they discarded it. I went down to the shop and asked to see the material. What they handed me was something that UFO researchers had been looking for since the mid-1980s. The August 3, 2000 cover letter was from Sarah Demb, Museum Archivist at the Peabody Museum.

Here was one page of an eight-page letter written in December of 1947, from Herbert Dick to J.O. Brew, a director at the Peabody Museum and one of his academic advisors.[3] In it, he described his survey trip to Bat Cave and specifically alluded to leaving for the Plains of San Augustin from Albuquerque on July 1, 1947. There, the party had begun digging exploratory trenches and mapping the cave and surrounding area. Furthermore, he said he and his party were at the location for two weeks... I thought, *So they were there! What a surprise!* Here was an entire group of archaeologists who had claimed they were not at Bat Cave in early July, when they actually were. Why? We asked Harvard's Peabody Museum for permission to publish relevant parts of a letter that came in a packet dated December 14, 1947.

The excerpt printed below appears on page 3, paragraph 3 of one page of some material sent to the author by Sarah Demb, Peabody Museum Archivist of Harvard University.

> ***Our party proceeded to the Plains of San Augustin on July 1, 1947.*** *We went immediately to the "Y" Ranch, owned by the Hubbell Cattle Company, where I asked Mr. James L. Hubbell, president of that company, if any archaeologist had been in the area in the past year. He stated that he had seen several strange cars on his land but they had never stopped to see him. I asked his permission to run exploratory trenches in Bat Cave and explained my plans in detail. He readily granted this permission and we established a camp at the foot of the cave and began trenching.*
> 
> ***We spent two weeks carefully contouring, mapping, and trenching the cave.*** *The trenching proved significant and the results were better than we had hoped for. The projectiles from the lowest levels were similar to those collected by the aforementioned people from a blow-out site west of the cave and across the old lake basin from it. Until that time none of the material collected in the cave itself had given any indication that it might have been occupied by Early Man.*

**THE CONFLICT**

It was obvious that the Dick letter found in 2000 trumped all major UFO investigators in the early 1990s and had a distinct bearing on UFO publications and outlooks in these years. This included the above-mentioned report, *The Plains of San Augustin Controversy,* July 1947.[4] One just does not find a 50-year-old detailed report in such files for no reason. Why was the letter in the files? It was clear to me that Dick had been asked to write it to offset another graduate student's claim for the right to do graduate research at the Bat Cave site.

On the other side of the conflict was Wesley R. Hurt, a former colleague of Dick at the University of New Mexico (UNM). Hurt was also a veteran and had served in U.S. Army counter-intelligence during WWII. He was some three years older than Dick and had a very forceful and sometimes abrasive personality. This may have been one of the reasons Dick transferred from UNM to Harvard in 1946. It became obvious from the correspondence in the Harvard/Peabody files that Dick and Hurt had competing plans for the Plains of San Augustin and Bat Cave area.

Essentially Hurt and a friend, Daniel McKnight visited the Bat Cave site during Christmas vacation of 1946. An important part of the protocol for initially choosing a dissertation site in field anthropology was to choose a topic or specific site and discuss it with your advisors or Ph.D. committee. If they agreed with the plan and no one else had laid claim to the site,

*Figure 10.2.* ***Dr. Wesley R. Hurt,*** *1919-1998, was a well-known and respected archaeologist and professor. He authored numerous articles on native Americans. His graduate work took place at U. of Michigan, UNM and U. of Chicago. He spent many years teaching at Indiana U. Photo courtesy of Indiana U. archives.*

then a paper or announcement was published, declaring intent to investigate it. According to a letter written by Hurt to Brew on August 20, 1947, he and McKnight were working on a preliminary report, which would be completed in September of 1947, when Hurt got back to New Mexico from a summer job in Alabama.[5]

In Hurt's August 1, 1947 letter to Herbert Dick ten days later, he said he had made reports to several museums, some influential people in the anthropology field and a publication called *American Antiquity*.[6] He was clearly playing both ends against the middle. He told Dr. Brew he would be working on a report in September and he told Dick he had made reports to prominent people and an anthropology publication. Dick showed Hurt's letter to one or two people at UNM and was told since nothing was published, he should not write to Hurt but continue his work on the Plains of San Augustin. Hurt apparently did not understand at the time that to claim an area for his own research, he was to declare his intent in something published, or by a dated letter from a publisher that he has submitted material. *American Antiquity* did publish a very nice piece of research on the Plains of San Augustin in 1949 done by Wesley Hurt and Daniel McKnight. It is doubtful it could have been finished and submitted before January 1948, five months after Dick's

Bat Cave survey. Dick was advised at UNM and no doubt at Harvard of this publishing protocol. It is believed he published his intentions in an anthropology newsletter at Harvard in the spring of 1947. Bat Cave was his. The mistake Hurt made was to assume his abrasive personality and age would hold the Bat Cave site until he was ready. This intimidating way of proceeding may have worked in the anthropology department at UNM, but was of little consequence in a traditional institution like Harvard.

Dick got busy in the Winter/Spring of 1947 and filed the proper papers, found funding, wrote the land owner, and arranged the stratographic survey to begin from July 1 to July 14, 1947. Dick got the land owner James L. Hubbell to agree to give him exclusive rights at the Bat Cave site.[7]

Hurt, upon his return from Alabama in September of 1947, had heard Dick had been to the cave site but apparently did not learn of the two week stratographic survey a month and a half earlier until he was told by Hubbell. To add insult to injury, Hubbell expressed some remorse at having given Dick exclusive rights to the cave through the 1948 season. The Dick and Hurt communication with mutual parties was telling. Dick understood one thing from one party; Hurt understood something else. Communication had clearly broken down. In a September 30, 1947 letter to Dr. Brew, Hurt pathetically still did not get the picture when he indicated Dick had "the understanding that he was to obtain my permission if he did more than look over the sites."[8] Dick had come and gone a month and a half earlier and was on his way to becoming famous.

We believe that it was this completely irrational tone of Hurt's that prompted Dr. Brew to ask for the letter from Herbert Dick to protect both Dick and Harvard in case the conflict escalated, which it did. In a Peabody Museum newsletter in the fall of 1947, a very pleased J.O. Brew referred to Mr. Dick's dig at Bat Cave and studies of early man for the Peabody Upper Gila expedition. Dr. Brew became director of the Peabody Museum in 1948. Dick was protected now by the prestigious Harvard umbrella and his association with Brew.[9] About 1935–36, the Peabody Museum at Harvard began some exploring and research on a huge part of southwestern New Mexico, designated as the Upper Gila expedition. Bat Cave was in this area. It is believed they were anxious to do some archaeology in the area. The coming of WWII delayed Harvard's Peabody plans to sponsor archaeological digs there. Dick's arrival at Harvard in 1946 and his knowledge of southwestern archaeology was to work out well for them and their desire to expand on the ground research into the Upper Gila area.

John Otis Brew retired in 1967, a year after the Peabody Museum's 100th anniversary. He passed away in 1988. Apparently someone going through his correspondence ran onto the December 1947 letter from Herbert Dick. This letter was filed in the archives at the Peabody Museum where it was found in the year 2000.

As they made camp that warm July evening, Herbert Dick must have thought back through his relationships at UNM and his first year at Harvard. It had been quite a year. Late in 1946 he had married a very nice girl from Tulsa, Oklahoma, Martha Morris. She was with him now and so was her brother Jeff. Good friends Harold Brown and his wife and Gordon Carter were also helping set up the camp in front of the huge, yawning Bat Cave, on a rise about 100 yards away. Martha's dad Alex Morris with surveying experience and equipment was due in the next day or so to run some surveys of the site for the stratographic survey. If all went well they would be done in two weeks; that's what he told James Hubbell the land owner on his way in.

## A METEOR?

The significance of this letter in Dick's file was enormous. We had already started to find strange, unearthly material at the arroyo crash site, but finding the archaeologist and his party were camped only ten miles or so away on July 2 gave considerable credence to Barney Barnett's story. It was much easier to see the scenario now. The archaeologists arrived in late afternoon and set up camp. No doubt, a fire was built sometime in the evening. We believe one or more saw the pulsating orange light as it came to earth and extinguished itself behind the low scrub hills in the sandy arroyo across the Plains.

There was a particular sensitivity for scientists in New Mexico in those days concerning meteorites. Several had crashed around the world, and they were presently in the news, especially big ones. There was another reason for meteors to be noticed and tracked in this part of New Mexico. Southeast of the archaeologists' camp, over the Sierra Cuchillo mountains, 30 miles away, was a small town known as Hot Springs (called Truth or Consequences since 1950). To the south was the Chihuahuan Desert, where hundreds of meteorites had been found.

It was peaceful out there and things had gone well. He and Martha had met the Browns and Gordon Carter coming up from Texas in mid-afternoon. At Socorro they had bought a big chunk of ice that was now in the wash tub in the commissary tent cooling the perishables with a burlap sack on top. It was peaceful out here away from the hustle and bustle of Cambridge and the slower paced Albuquerque. Off in the distance as the sun headed behind the San Francisco Mountains, could be heard the call of a horned meadow lark somewhere in the sage brush... and a little later an answer. The sun now only a memory had disappeared two hours or so ago and the chili and fresh biscuits Amelia Brown had brought up from Texas had been a real hit with the party.

It had been a good meal. Everyone was tired but too excited to sleep. Now and then could be seen a flashlight shining inside one of the tents casting weird shadows as one of the girls bustled around adjusting the bedding or a sleeping bag on one of the canvas cots. Another round of the coffee pot had just taken place as the moon was making its appearance rising behind the San Francisco Mountains. One of the men stood up and pointed north towards the Gallinas Mountains and said, "What the ...Hell...is that?" Not everyone could see and he said there, there. It seemed to be a dim, orange light with an occasional brighter pulse very low, headed West. By now, someone over near the large wall tent said, "I see it... It's going away. It's getting dimmer." Then someone said "I see it there below the mountain...no... no it's turning."

Momentarily it was lost behind some lower hills, then when it reappeared much brighter someone said, "Look how low it is..." Another voice chimed in, "It's going to crash." Then the light went out. There was some discussion about a plane or maybe a meteor... in the darkness, a man's voice said

"Meteorites don't turn..." Then another voice... "I think it was too fast for a plane...Did anyone hear any noise?" Silence... then one of the men said I'll lay this ax handle pointing that direction. Maybe in the morning we can see something. A girl's voice inside a tent was heard, "Hope no one is hurt." One of the men with military experience said quietly to a couple of the men still around the glowing embers of the fire, "If it was a plane, it was low in gas...I thought I saw a flash after it hit... Was there an explosion?"... Another male voice, "I didn't hear anything...couldn't be survivors." After everyone turned in there was still mumbling in the tents, then silence as a slight breeze caused the last log on the fire to glow and shift slightly — then all was dark; all was quiet. It had been a long day.

If the bright light across the Plains had been a meteorite, it would have been much higher and would have come from the northwest, and then continued in a southeasterly direction over the archaeologists' camp at Bat Cave and quite possibly have crashed where many others had been found thirty or so miles to the southeast. If it was a plane, why was there no noise? If it was a large meteor, it could be worth hundreds of thousands of dollars. Harold Brown was checking out the surveyor's transit earlier in the evening. Perhaps they could take a rough reading off the transit compass ring and get a general bearing in the morning.

The old brass surveyor's compass Dick was known to have carried in the field was still in the wooden box with the leather strap handle. It had line of sight brass arms at right angles to the huge 7-inch compass face. It, of course, was unusable at night. It would be very valuable in the morning, once daylight came, to sight a landmark in the far distance. Would the roads they had seen coming in take them over in that direction? Would it get them over to the other side of the Plains? An event of this magnitude would certainly have delayed the archaeologists' work until a party of them could cross the Plains and see what the source of the light was.

It would have taken an hour or so, driving over unfamiliar ground and roads to reach the other side of the Plains, ten miles away. When they finally got close to the arroyo where the light was extinguished, it was in a depressed area somewhat lower than the surrounding ground. Whatever the object was, it was in that arroyo out of sight. About a quarter-mile east of the crash site was a sturdy drift fence with no gate for a mile or so in either direction. Since they were almost to the arroyo, it would be prudent to park the car, crawl through the fence, and walk down to the where they thought the crashed object had come to rest.

It must have been quite a surprise to Barney Barnett as he was looking at the crashed craft and bodies to see four or five people coming from what he told Vern Maltais was the "opposite direction." He probably knew that they were connected with an archaeological dig, especially the way they were dressed. It must have been a cautious meeting at first, each party trying to ascertain what the other knew and what should be done.

We have no idea what the conversation might have entailed, but there was certainly plenty to look at and probably no shortage of questions about the origin of the craft and bodies. They probably had less than a half-hour to gather at the site before the military arrived in what we believe was one or two large trucks and a small vehicle, such as a Jeep accompanying them.

The 1947 Chaco Canyon Conference was held at the end of July and beginning of August that year at the visitors center 150 miles or so north of the Plains. It was the first conference after WWII, and hosted many archaeologists from a number of colleges throughout the United States. Herbert Dick was there and made a small presentation about the work he was just beginning. A rumor made the rounds that a group of archaeologists traced a bright light to a crashed spaceship with little bodies. The rumor must have been from someone in the Dick party talking to someone going to the conference. Dick, no doubt, was annoyed and amused, and probably acted as surprised as anyone when he heard about the "archaeologists discovering the spaceship." Eventually this rumor was discounted.

Dick actually had two interlocking secrets to keep quiet. He first wanted to secure the Bat Caves site for his Ph.D. project. But he did not want it known that he was there at the beginning of the month doing his stratographic survey. Dick especially wanted to keep his work quiet, as Brew had suggested. Dick started on July 1st while Hurt was finishing up some field work in Alabama.

Wesley Hurt was claiming this site for himself, but as far as we could tell, or anyone at Harvard could ascertain, Hurt had generally not gone through any academic process or published his intentions. Harvard advisors had encouraged Dick to get into the site and get the survey done while Hurt was away. Dr. Brew at Peabody especially supported Dick in the promising site tied in with Peabody's Upper Gila research that Peabody had started before the war.

Then to actually have come upon a spaceship with bodies meant that he and his party had to keep both secrets. It would seem a little unethical for Dick, with Harvard support, to take the Bat Cave site, so all agreed to keep the early July 1st beginning at the site quiet. As it turned out, looking at Hurt's letters in Dick's Peabody file from August and September of that year, that he was brash and unreasonable, and apparently was trying to keep others from the site through intimidation. Dick's advisors also wanted to see if any of Hurt's claims to the site would surface in any archaeological literature that summer — none did. Harvard/Peabody and Dr. Brew especially wanted to not have open conflict with Hurt's latest school, the University of Chicago. The strategy worked. Dick got in and did his survey that July of 1947 and came back in 1948 and did a classic dig, and was famous for what he found. Wesley Hurt transferred to the University of Michigan where he received his doctorate for research in another area. He was very bitter about the whole experience and never forgot it.

Dick indicated to the UFO researchers who contacted him in the '80s and '90s that he had been on the Plains later, but would not admit being there the first week or ten days of July that summer when the Roswell craft was discovered.

The Barney Barnett story of the Plains discovery surfaced a little later than the main breaking Jesse Marcel story of the Roswell events. Although no date was initially given to the Plains crash, Barnett's story with the small bodies seemed to be a natural part of one or more space craft crashing on Earth. This was amply conveyed in the Berlitz and Moore

1980 book, *The Roswell Incident*. At that time it seemed only logical that verification of Barnett's story hinged on finding the archaeologists. They were and had been in deep cover for many years.

When one or more of the archaeologists failed to appear after a decade of searching, the two stories started to drift apart. To complicate matters, various ex-military people and some civilians started claiming involvement or knowledge of the Roswell event, especially after all the worldwide publicity it received. Volumes of material have been written on those coming forward to get a share of the Roswell publicity. Occasionally someone would mention a connection with the Plains crash, but when Dick convinced all of the main researchers he and his party were not on the Plains in early July, the arroyo crash site and Barney Barnett's stories were pushed to the back burner and all but forgotten. One researcher, Stanton Friedman (author of *Crash at Corona*), maintained there were two crashes that first week of July 1947. Friedman was the only major researcher who thought that the Roswell crash and Plains crash happened about the same time. The arroyo crash site originally gained attention when Gerald Anderson came forward after a 1990 *Unsolved Mysteries* TV program. Friedman supported Gerald Anderson until a conflict between Anderson and other UFO writers heated up in 1992. Then he backed off, but through the bitter conflict Friedman maintained there was a strong possibility of a Plains crash. He was right!

The military secrecy lid placed on those found at the crash site that morning, no doubt would have been a welcome restraint to keep others in the Dick party from saying anything. The silence about the camp at Bat Cave and the discovery of the crashed flying disk worked to Dick's advantage, although someone had obviously been responsible for the rumor at the Chaco Conference. As far as we know, Dick kept his early arrival at Bat Cave secret from other archaeology students, and he certainly had no motivation to talk about the UFO crash either.

The Harvard/Peabody letter in the archives, of course, does not place him at the arroyo crash site with Barney Barnett and the military. However, his Bat Cave camp's proximity and dates he and his party were there ten miles away from the crash site increases the chances greatly of his possible involvement. We know much about this man today whose letter to his advisor over fifty years ago changed the direction and emphasis of the Plains of San Augustin investigation, and parts of the Roswell investigation as well.

Who was Herbert W. Dick? Hailing from Illinois, he was born in Chicago in October of 1920. Health problems prompted his father to bring his family west in 1921. The elder Dick had contracted tuberculosis in WWI and had been advised by his doctors to leave the Chicago area. Herbert's father, an electrician by trade, then took work in the mines near Vandium, New Mexico. Due to problems in the mining industry created by the Great Depression, the family moved to Albuquerque in 1932. According to some old Albuquerque city directories, his father, also named Herbert Dick, took a job as an electrician with the Veterans' Administration. He was located there from 1934 through 1936. Herb, as his friends called him, attended school there and graduated with the Albuquerque High School Class of 1939. He also began classes at the University of New Mexico that year.

He was enrolled in college when WWII broke out, and spent most of 1942 as a fire spotter for the U.S. Forest Service in the Gila (pronounced "He-la") Primitive Area in southwestern New Mexico. He then became part of an archaeological crew excavating at Chaco Canyon. A short time later he joined the Army Air Corp and engaged in combat intelligence work in England until 1945. He then returned to UNM where he received a bachelor's degree. In the summer of 1946, he enrolled in Harvard to work on his master's degree. From the fall of 1947 to 1950, he was in graduate school at Harvard while also working toward his Ph.D., which he completed in 1955.

*The Harvard Gazette*, the Harvard University general catalogue of course offerings of September 20, 1947, listed the fall registration calendar of events. Herbert Dick was probably in Cambridge by Monday September 22, which was the registration day for "new graduate students and men returning to the School of Arts of Sciences." [10] (In those days the adjoining Radcliffe College was an affiliated institution for women.) In 1946 Dick was listed in an arts and sciences directory as holding an A.B. degree from UNM, and was in his second year at Harvard graduate school in the anthropology department. Indications are that he was just finishing up his master's program. It is believed he was also a graduate assistant for the 1947–48 school year. As was the practice among graduate students, Dick shared some instructional duties with Dr. Brew and one or two other anthropology professors that school year. He did not begin his full-time college teaching until the school year of 1953-54, and taught until his retirement in 1980. He had a distinguished college teaching record.

One of the factors that worked against UFO investigators was their relative youth. For literally years, these investigators spun their wheels looking for a university professor with young students. In reality, the Dick party consisted of a group of mature adults in their mid to late twenties. Dick, in 1947, was 27. The fact that these students were older was due to the disruption of educational plans during World War II. Many returned to school in post-war days. For the next five or six years, these mature veteran students did not need as much field supervision as the younger students did before or after the 1950s.

There may be other good reasons for the Bat Cave timeline cover-up. Instead of the denials the Dick party gave originally that were intended to place them farther away from the crash site and timeline, the chances now are much greater that they were the archaeologists that Barney reported seeing at the site. If Dr. Brew had not asked for a statement from Dick for the files regarding the Hurt conflict, then the Dick party denials would stand to this day.

One of the greatest mysteries of this event is why Barnett did not mention others at the crash site in addition to the archaeologists. We do not know how much time, if any, the civilians had to compare notes and introduce themselves. We do know the military showed up soon after. Historic evidence indicates there was a party from the Anderson family, including Gerald and some other relatives at the site while Barnett was there. If Barney knew about the Andersons, why did he not mention them too or give reference to the family when he told others about the crash site? The following may have some relevance.

Barney Barnett had worked hard and bided his time to

get a promotion to what must have been a prestigious job with the USDA, Soil Conservation Service. It was about as good a job near a reasonably fair-sized town, Socorro, where he lived, as there was in New Mexico for a civil engineer in this line of work. He transferred to the soil conservation service in May of 1946. He had only been on the new job a little over 14 months when he discovered the crashed UFO. Conservation people, especially from the government were not as readily accepted in those days as they are today. Public relations and acceptance of men like Barney in the field, dealing with ranchers, was extremely important. Barnett may have considered several options regarding who he said was at the site. It is possible that Barney with the advice from Fleck Danley decided to leave any local New Mexico people out of his recollections (to those he did tell). The Anderson family had relatives in Magdalena where Barney was based. The Andersons visiting the site were from Albuquerque where the USDA had an office. Any mention in New Mexico papers of this event could damage the credibility of the Soil Conservation Service and the USDA, and ultimately might affect Barney's reputation and job.

Initially, we think only his wife Ruth and boss Fleck Danley knew. Barney may have been hedging his bets by only telling others about the crash who lived out of the area. He told Alice Knight and her husband J.B. at Thanksgiving time in 1947. They were from Dalhart, Texas. The next people we knew he talked to were Vern and Jean Maltais from Bemidji, Minnesota, in 1950. After his 1959 retirement, we know he did tell a neighbor, Harold Baca, sometime in the mid-1960s.

There was a hidden fork in the road here for UFO investigators between 1980 and 2000. If it had been known in those days that the Dick party had been on the Plains in early July doing their stratographic survey and they had been truthful about it instead of covering it up for many years, Barney's Plains story might have been investigated more thoroughly. He and his alien body descriptions would probably not have been moved 160 miles or so to the east to accommodate the per-conceived Roswell scenarios and ludicrous theories. This probably would not have affected the Roswell investigation a great deal, but a two-crash, two-UFO scenario would have brought the Plains crash and Barney's story under a much different light.

Herbert Dick was a gentle, kind man with a good sense of humor. In his later years he could have been anyone's husband, older brother or father. He became the forerunner of research into Paleo-Indian agriculture. Herbert Dick inspired thousands of students. Those who were interviewed thought he was an excellent teacher. Over the years, a number of his students also became teachers, and many continued to obtain their Ph.D. One of his former student associates at the University of Colorado Museum thought that he was a very perceptive person when he suggested she might concentrate more on her studies, and less on boys! Dick was also a man who helped orchestrate the illusion that convinced the determined, and somewhat persistent, UFO investigators that he and his party were not where they actually were.

Herbert Dick passed away in the early 1990s, in a state he loved and gave so much to. In the final analysis, we should ask ourselves, who among us with a college background or advanced degree would embarrass a prestigious university or jeopardize a chance to acquire a Ph.D. then or now? In retrospect, if the Dick party was involved, it was a very wise decision not to make public a flying saucer story that ultimately the government would deny or cover up, and few would believe.

# CHAPTER 10
## REFERENCES

1 *The Plains of San Augustin Controversy, July 1947: Gerald Anderson, Barney Barnett and the Archaeologists*, prepared by George M. Eberhart. Fund For UFO Research, Chicago, Illinois, p. 3
2 Ibid., p. 9
3 Dick, Herbert W. Letters to Dr. J.O. Brew, Dec. 14, 1947, Accession# 48-44 Upper Gila Reconnaissance Records _ Bat Cave Series: Expedition Correspondence 1947-1953, Box 1.2.
4 Op cit. *San Augustin Controversy*
5 Ibid. Hurt, Wesley R.,Correspondence with J.O. Brew & Herbert W. Dick.
6 Ibid. Hurt correspondence
7 Ibid. Hubbel, James L., Letter to Dick, Sept. 14, 1947
8 Ibid. Hurt, Wesley R. Letter to J. Brew, Sept. 30, 1947.
9 Ibid. Brew, J.O., Published material in the *Peabody Museum Newsletter,* dated Fall 1947.
10 *The Harvard Gazette*, Sept. 20, 1947, Vol. XIII, No. 1.

Note: Some of the background on Herbert W. Dick's personal life came from numerous obituaries found in college archives and correspondence with his wife Martha Moris Dick.

# CHAPTER 11

## BAT CAVE AND BEYOND

It has been a long time since they bladed this road, I mumbled to myself as my left front tire hit the low side of a huge pothole. I was on my way to Bat Cave on a hot and somewhat windy July afternoon. I could look northeast and see the mystical Plains shimmering through several mirages and areas of blowing dust. Now, though, I had to concentrate on this road. The last rain some months before had overfilled the uphill ditch, and water had spilled onto the road, carving narrow troughs as it ran across. They were especially bad in the low spots.

I was proceeding south on Forest Road 28, often called the Bursum Road by the locals, and had just passed the Long Canyon Road leading east over the Tularosa Mountains. The high ground directly ahead of me, some ten miles away, was the O Bar O Mountain. The man I had met out near Highway 12 was from the Socorro Electric Co-op out of Magdalena. From his map we were able to determine that I wanted to turn on the next road ahead towards the Plains.

I found it, but it was not much better than the one I was on, although it did offer some relief from the ruts and potholes going across the road. Now I could drive down them, but this novelty soon wore off. In a mile or so I eased onto the Plains and the road flattened out some. In the distance I could see that it led to a house. I was hoping to find someone home. As I got closer, I could see the top of a white pickup truck. The bottom, being like all other vehicles in the area, was a sort of caked-on light muddy brown.

It was an octagon-shaped stone house with a very thin wisp of smoke coming out of the chimney on the back side. I had somehow mollified the dogs who usually take exception to visitors. I knocked on the door and a rather surprised, but pleasant man opened it. I asked if this was the way into Bat Cave and could I go in. His answer was affirmative to both. He looked out at my pickup and asked, "Ya takin' in that rig?" I replied that I had planned to. He said nothing, but told me he and his boy, who was finishing up some lunch at a large round table, were about to leave. I told him where I was from and that I wanted to take some pictures. He said a group of students from Stanford had been in there about a month earlier and it should be all right. He asked if I had food with me, and I explained that I had some, as well as a sleeping set up... but I was not going to camp, just in and out by later that afternoon.

Western people are very quiet and reserved, and seldom talk in frank terms, but I could see something concerned him and he finally said, "We don't expect any rain, but if it does you probably couldn't get out until this time tomorrow. Tell ya what ya do," as he gestured with a weathered thumb towards a fence. "See them boards over there? Stand one of them up by the gate when ya go in. When ya come back this-a-way, take 'er down. That way we'll know ya got out all right." It sounded like a good plan to me and after assuring him all gates would be left the way I had found them, I started out. I felt very secure when I went in that a rancher was looking out for me, until I came out later in the day and found a note on the ranch house door to someone about feeding the horses. It read in effect, "My son and I will be gone several days. Hay in barn." So much for security!

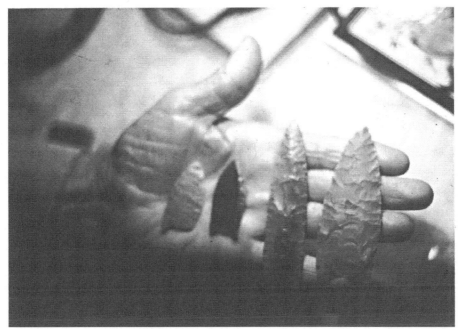

*Figure 11.1. **Projectile points** such as these were used by early Plains inhabitants. The smaller Clovis points date from 9,000 B.C. The large points are Sandia II. They date from about 11,000 years ago. Both point types went on spear shafts some 1,500 years before the bow and arrow was invented. Larger points reflect larger game in the area. In 1947 Dick found the skull and horns of an extinct buffalo, much larger than we know today.*

In the 59-mile-long lake bed floor are three contiguous basins that are at several levels below sea level. U.S. Highway 60 crosses the highest part of the old lake floor known as the White Lake Basin, at 6,952 feet. The main two-thirds of the lake floor is below a sill (a strata of ancient volcano magma over 50 feet below the White Lake basin.) The next identified floor level is at the C-N Basin at 6,894 feet; and the third and lowest level, to the southeast, is known as the Horse Springs Basin at 6,775 feet[1]. All of these Plains of San Augustin low lake level basins are associated with archaeological sites.

Generally, where water soaks in on the lower Plains, there is some vegetation. There are also areas completely void of vegetation, areas known as playas. Their shapes range from circular or oval to narrow and elongated, from several hundred yards across to only a few feet. In Spanish, playa means "beach." The term usually refers to sandy depressed areas that remain dry except after rains.

Playas are formed as the mountain slopes drain down streams, like the one running through the aspen grove, mentioned before, and into the main trunks of arroyos. The process is part chemical and part physical. As the water moves downhill it carries a solution of salts and alkalis. Since there are no drainage outlets on the Plains and much of the basin underground consists of sand, silts, and adobe clay, the salts and alkalis percolate up from below and spread out across the Plains' surface, forming these playas.[2]

When roads traverse these playas, the surface material compacts about seven or eight inches below the surface, and two parallel ruts are formed. All roads or tracks that run through these playas or anywhere on the old lake bed are deeply rutted. When it rains, water settles in these ruts making travel difficult until it dries up. When we flew over the Bat Cave area, the playa edges were honeycombed with lush vegetation. It is this vegetation that has made grazing on the Plains productive for over one hundred years.

As I drove along in the two deep tracks of the road, I was reminded of a statement made by Robert Drake, a University of New Mexico anthropology student, to the effect that he had been to Bat Cave in September 1947 and had seen heavy equipment tracks there. Perhaps the memory of the deep ruts outlasted the reason for them being there in the first place. According to the Harvard letter written by Herbert Dick in December of 1947, Drake was accompanied by five individuals on this trip. He and his University of New Mexico friends visited the cave and a nearby blow-out site in February of 1947.[3] Drale's recollection differs by five months. I found all the roads on the east side of the Plains were deeply rutted due to the nature of the compression of the playa soil and clay worn down by vehicles. I was also reminded that an early motorist once said, "Choose your rut carefully; you'll be in 'er for many miles." There was only one set of ruts here, deeply compacted and straight as an arrow. (About a week later, I had the pickup on a hoist being serviced. The attendant commented on the undercoat being worn off and, "The pan has really been shined up.")

The road ran northeast past the glacial cut, La Jolla Canyon, and below several tiers of high volcanic bluffs that were pushed out by flowing lava some ten million years ago. Ahead I caught an occasional glimpse of Shaw Mountain. The bluffs to my right were at the extreme northern end of Pelona Mountain. To my immediate right, I could see sections of cliffs over 100 feet high. I was looking for a particular bluff that jutted out at right angles. I could see that bluff a mile or so ahead; when I got to it, Bat Cave was just beyond. I was unprepared for the size of the cave. It was immense.

The floor of the cave is above the old lake bed and playas about 100 feet. The climb was up a fifteen-degree slope, around sage and rabbit brush and over boulders that had fallen off the cliff face and outermost roof of the cave. I was carrying considerable camera equipment in a back pack and salesman's sample case, as well as a tripod. The interior of the cave had a sloping amphitheater-like floor, varying ten to fifteen degrees in places. There were four smaller caves to the left of the entrance, one or two going in twenty-five feet or more. Except where excavated, they were low with smoke-blackened ceilings, and I made no attempt to enter them. I did note that the entrances could be blocked off in cold weather, and with a fire they probably would have been quite cozy.

Since it had been surmised that Herbert Dick's

*Figure 11.2.* **Small caves,** *left of large entrance, could accommodate up to a dozen people. The Dick party found many artifacts here in these caves, including bison remains, a leather bag filled with beans, basketry, sandal remnants, and corn cobs. In earlier days, ground level was much lower. Blackened ceilings of small caves indicate Paleo-Indian use for thousands of years.*

*Figure 11.3. **Bat Cave** from outside. Access is difficult on private, posted land. Several smaller caves are to the left. Earliest known maize (corn) was found here during 1948 archaeological digging season. It is believed caves did not have permanent inhabitants, but may have been occupied during the growing season. Evidence of squash and beans were found here. The early Chiricahua Indians lived and hunted in this area. The cave, 57 feet deep, 66 feet wide, and 75 feet high, overlooks the Plains of San Augustin. Elevation 7,000 feet.*

archaeological stratigraphic survey party, working here in 1947, might have seen something come down somewhere on the Plains, I had brought the camera equipment to photograph possible angles. What I was able to determine was that inside the cave from the point of deepest possible penetration, an overview of the Plains existed for an 80-degree or 90-degree arc (Figure 11.3). This arc widened as I went towards the forward sections of the cave. The right-angle bluff on the southwestern edge of the cave made it impossible to see anything in that direction.

Looking out of the entrance to the cave, I could see surface haze, and other blowing dust, especially on a windy, hot day like this one. All of these conditions were not a factor in seeing onto the Plains from the higher vantage point of the cave. When I left the cave and walked lower down the slope, the bottom half of the hills on the other side of the lake bed was obscured by haze, dust, etc. The flat ground in front of the cave did offer very good camping terrain and had been used by Dick's party in 1947, 1948, and again in 1950. In Dick's *Bat Cave* (1965, p. 3), a photo of the 1948 campsite shows it to be on level ground in front of the cave.[4] Here, there is a 180-degree arc of visibility, except when various desert factors obstruct distance vision in the daytime. If a lighted object of any kind landed or impacted at night in the ten miles across the Plains, the twelve miles southwest towards the Bursum Road or some twenty miles to the northeast, it would have been visible in clear weather if there were no obstructing hills or terraces.

## OTHER ARCHAEOLOGISTS

Earlier in the Plains research before the analysis of the extraterrestrial metal foil shards, we researched other anthropology and archaeology work near the Plains in 1947. In my research, I sent for, through inter-library loan, numerous articles written by scientists from 1925 to 1965. These earlier articles gave me a perspective about what work had been done, not only on the Plains, but in the surrounding area. These included water reconnaissance, geological features of the Plains, and radiocarbon dating of campsite samples, as well as numerous other scientific reports.

I zeroed in on reports dated 1948 or later. Scientific work may have been conducted in 1946 and 1947 in the field, but we found no reports dated the year of the field work, and with many the following academic year. It is the nature of the anthropological/archaeological researchers to visit a site, make a report, do a preliminary dig, collect some samples, and apply for funding, before going back to complete the work.

I would like to also note that a site may have a crew of scientists supervising excavation, often students and/or some hired laborers, camp staff, and visitors. Associated with archaeological excavations, there are trips into town or the nearest store for supplemental groceries, drinking water and ice; laundry to be done, and gassing and servicing of vehicles. All of this movement would mean people associated with a dig coming and going from the camp every few days. The Pine Lawn Dig, some 40 miles south of the Plains, would have produced some traffic on New Mexico's Highway 12, the main north-south road in the area. (See map, Figure 9.3.) Anyone going to Reserve from the main New Mexico transportation and academic centers to the north would have used New Mexico 12 along the western side of the Plains to get to the Reserve area 40 miles south. Barney Barnett, we believe, first saw his "glinting metal" from this road as he drove to the Catron County seat south on NM 12.

While researching local brevities in the Datil section of the August 21, 1947, *Socorro Chieftain*, I ran onto the following article:

> Mr. Danson and Mr. Brew of the Peabody Institute of Boston, who have been studying ancient ruins and the customs and habits of the vanished dwellers visited Datil this week. Mrs. Cleveland took them to the old Indian cave in Thompson Canyon. They will return next summer and chart and study this part of New Mexico.[5]

We discussed Mr. Brew previously as Herbert Dick's advisor at Harvard/Peabody. It was his desire to re-establish Harvard's upper Gila project, started before WW II. When Herbert Dick showed up at Harvard in 1946 (he may have been recruited), Brew and the anthropology department probably thought Dick had experience and knowledge of the project area, with a desire to do graduate work there. Harvard would gain some degree of prestige with a viable project in the area. Possibly one of the reasons Dick's presence at Bat Cave in early July was kept confidential was that he had received a very quiet stipend of $800 from the university to help finance the stratigraphic survey. We are not sure how ethical this was in those days, but Dick's friends probably knew that he lacked funds and a sudden influx of money might call attention to Harvard University and/or Peabody Museum.

I wrote the Peabody Museum at Harvard University and was able to learn that Dr. Brew was the curator of North American Archeology at the Museum from 1941 to 1947. Edward Danson was a student at the University of Arizona hired to accompany Brew. He later undertook graduate work at Harvard in 1948.[6]

Danson and Brew traveled some 6,200 miles during the summer of 1947. It was more than likely that most of the 6,200 miles included a round trip to Cambridge, Massachusetts. The actual area they surveyed was some 30,000 square miles. They began what they called a rapid "reconnaissance without digging." In the article from which some of this information was obtained, they did admit to "occasional excavations of the car" or getting stuck. The men began their trip in Clifton, Arizona, working their way north some 30 miles up the San Francisco River to the Blue River and into the headwaters of the Little Blue. They then went to Silver City, north to Diamond Creek in the Beaverhead District, northwest to the O Bar O Ranch and vicinity, and then to Reserve.[7] They spent some time in a triangle between Reserve and Springerville, Arizona, and Datil, New Mexico. This pretty well fits in with their visit to Agnes Morley Cleveland about the second week of August 1947. The Cleveland Ranch was located in an area between Datil and Pie Town called White House Canyon. The canyon was named for the former Morley log ranch house, which was painted white.

In 1947 Agnes Morley Cleveland was seventy-three years old. She was one of the most famous western women of her day. Born on a cattle ranch in 1874, she moved to the Datil country about 1880. Her father, an engineer, was killed in Mexico when she was about ten years old. Later she was able to expand her horizons when she attended a young ladies' finishing school in Philadelphia. Recalling the hardships of her youth on a large western ranch, she wrote *No Life For a Lady* (Houghton Mifflin, 1977). Her book describes the experiences and hardships endured by her mother, brother, sister, and herself. The book received very good reviews from the *Atlantic Monthly*, and the *New York Herald Tribune,* and the *New York Times.*

Brew and Danson divided the long trip into thirteen areas. The ninth of these areas is listed as Datil-San Augustin Plains. The Plains are mentioned several times in the narrative. Highway 60 is also mentioned, as are Reserve, Datil, and Magdalena.[8] We do not know if Brew and Danson visited the Bat Cave site while the Dick Party was surveying there. They did mention in their *El Palacio* article, that Dr. Kirk Brian, a noted geologist from Harvard, was there while the stratigraphic tests of the site were being completed by Dick. That would be towards the end of the second week in July 1947. We have no direct information how much time Brew and Danson spent on the Plains sites, but mention of towns close to the Plains would certainly indicate they traveled roads surrounding it. Under the circumstances, considering the pending friction between Dick and Wesley Hurt, Brew may have suggested that Dick get into Bat Cave early in the summer of 1947. However, Dick's presence at Bat Cave and the exact date was kept quiet for the time being.

It is thought that Dr. Brew and Danson's travel in the area at about the same time Dick was at Bat Cave might give Dick an advantage if friction between the universities ever arose over the Bat Cave site.

Probably some of the most complete and interesting work done of the entire Plains of San Augustin Basin was done by Wesley R. Hurt, Jr., previously mentioned, and Daniel McKnight. They surveyed the terraces and old lake bottom for signs of early human habitation. The terraces around the extinct Lake San Augustin were formed by wind and wave action over thousands of years. As the glaciers that filled in the old melt water lake began to recede about 18,000 years ago, terraces appeared along the old shore lines. The ground above the terraces was relatively flat, and some of these terraces were mapped by Hurt and McKnight as camp sites of the early people. Also associated with the terraces were bars and beaches. It was from a high terrace on an old island that I first saw the gap, going through the buck brush, in the arroyo.

This was no doubt the work that Hurt and McKnight were doing that Hurt used as "previous work on the Plains" to justify his start on his Plains research. Hurts' correspondence of 1947 lists his visiting the Plains and Bat Cave during Christmas vacation of 1946. We knew Hurt was "putting something together" in the fall of 1947, especially after he learned Dick had been there doing his survey. Dick is quoted in his December 1947 letter to Dr. Brew as saying, "Mr. Hurt's immense interest in Bat Cave itself apparently arose after he had been informed of the work I had done there."[9] This may or may not have been true, but Hurt and McKnight spent no or little time on the Plains in 1946 or '47 as both were full-time students with summer jobs. It should be pointed out that when the "monsoon rains" came, usually in late September, the Plains were essentially closed for research.

On the south side of the Plains some old volcanic flows end in series of bluffs hundreds of feet high.[10] The wind and waves worked against these bluffs while the lake was at its highest, some 7,075 feet above sea level. Erosion worked on softer stratas of these bluff undercuts, and caves began to appear. The most prominent of these caves is Bat Cave, in what geologists call a wave-cut cliff.

As the lake gradually receded, different wave-cut terraces were exposed. Over thousands of years as the lake shrank, the camps were moved to lower terraces to be near the water. Eventually the water receded into the flat areas of the old lake bed, and the camp sites followed.

Hurt and McKnight did some excellent work on the Plains, published in *American Antiquity* in 1949.[11] The rather late published material had no impact on Dick's initial dig in 1947 and later dig in 1948. If it had been published two years earlier, it would have given Hurt an edge in the conflict.

Hurt and McKnight identified twelve archaeological sites on the terraces around and in the old lake bed.[12] As some of the terraces and lake bed sites were abandoned, whatever archaeological material that was on the surface or buried was exposed as the over-burden worked its way down by erosive process to a more non-erosive surface. These surfaces, where many artifacts were eventually found, are known as hardpan. Sites such as this, which have been exposed to harsh winds and other elements, are known as blowouts. These terraces and lake bottom sites revealed various scrapers, projectile points, whole or in part, other kinds of worked flakes and pottery shards, as well as the milling and grinding stones known as manos and metates.

Hurt and McKnight in their research paper said that in the 1930s, at a site two miles north of U.S. Highway 60, a human skeleton of an adolescent female was found.[13] The group from the University of New Mexico, who excavated the site, thought she might be associated with a people from the Folsom Complex about 11,000 years ago. Hurt and McKnight, as well as other leading scientists, felt that the first people to enter the San Augustin Plains may have been from this group. Near the girl's skeleton was found a Folsom point, which was made to attach to a spear shaft. The Folsom people who hunted this way, may have followed big game into the Plains area after the lake receded. People of this time period are generally referred to as Paleo-Indians. Apparently, the camp site was used up to the more recent Hohokam period from 1000 B.C. to 1300 A.D. It was not until about 1000 BC that the bow and arrow began to be used in the southwest. Other pieces of Folsom points were found at the C-N site located at the northeast end of the Plains. An example of Clovis and Sandia II points can be found in (Figure 11.1).

Hurt and McKnight located some of their sites through locals who hunted and collected artifacts. In a very comprehensive article written in *American Antiquity*, Hurt and McKnight acknowledged the help of two Magdalena men, Jack DeVaux and Lee Garner. In 1998 the author was able to contact Wesley R. Hurt, living in Albuquerque, who at one time had been a college professor at Indiana University. He had lost track of Daniel McKnight and the others he had worked with when surveying the Plains. I asked Hurt how often he was able to do his survey work and what sites were surveyed and when. He replied he could no longer remember what sites were worked in those years.

I was able to locate a mention of a Mr. Hurt in Ruth Barnett's diary entry of August 20 and possibly 21, 1947.[14] I asked Wesley Hurt about this, and he said in 1947 or 1948 he had gone with "a government man" to the Plains but did not recall his name.

The Pine Lawn Valley is seven miles west of Reserve, New Mexico, outside the western edge of Salado SWCD boundary on New Mexico Highway 12, (See map, Figure 9.3). Several archaeological surveys have been carried out in this area since 1907. Dr. John B. Rinaldo was sent out by the Chicago Natural History Museum in April of 1947 to conduct a more thorough survey of the area. Rinaldo surveyed sites within a 25 mile radius of the Reserve area. The survey covered 100 square miles where more than 100 archaeological sites were located. We believe that a friendly rivalry between Rinaldo's Chicago Natural History Museum and Harvard's Peabody Museum may have stimulated an interest by Peabody to restart archaeological work on the Plains area 40 miles to the north.

Rinaldo's survey was preliminary to intensive digging, which began later in the summer of 1947. Sites surveyed included: eight pueblos (stone or adobe house remains), some twelve pit houses (shallow circular or oval pits), which had once been roofed over, and numerous cave sites. The excavations were centered in the Pine Lawn Valley where some of the promising sites were located. The excavations were directed towards solving several problems that concerned the origin and development of a culture called Mogollon (pronounced mo-gee-on) from 1500 B.C. to 1300 A.D.

Although no specific timelines were mentioned, it is assumed the exploratory expedition and the digging phase in 1947 consumed some five months. Digs at the same sites in 1948, 1949, 1950, and 1951, probably lasted two-and-one-half to three months. One interesting comment concerning a spring in Wet Leggett Canyon indicated that during the dry July of 1947, the spring flowed uninterruptedly. This would indicate the archaeological work started in the spring and continued through the month of July. These and weather and ground water factors may have influenced Dr. Brew to recommend that Dick get an early start on his Bat Cave survey. Area weather records indicate a very dry summer that year with rains starting in September.

What helped make the archaeological work at the Pine Lawn Valley so productive was its proximity to the town of Reserve where the team of six or seven professional people stayed at the Pine Lawn tourist camp. They thanked a Mrs. Mary Crackel, proprietress, for the use of "electric current, water, and milk." A woman named Martha Perry was employed as a cook.

The dig at Pine Lawn was an archaeologist's dream as far as camping went. Hot showers at night, cabins instead of tents, no food preparation. The only work for the archaeologists was at the site, which is what they came for. Before I learned of the dates of the Dick party at Bat Cave, I really thought that one of the numerous students, about 12 in all, may have been driving on Highway 12 and came upon the crashed UFO, which we later learned Barney Barnett had found. The week of July 2 was only two days before the 4th of July holiday, and there would have been some traffic on Highway 12 to or from the Pine Lawn dig. This would also include boyfriends and girlfriends. A small party need not have an archaeologist with them to leave the impression that they were doing archaeological work, which would be true especially if they had to justify their presence to the Army. Even the laborers at Pine Lawn and Dick's helpers at Bat Cave were doing archaeological work, but some were not necessarily archaeologists. Archaeologists Martin, Rinaldo, and Anteves were thoroughly questioned by early UFO investigators. No indications from anything they said led to anyone in their party finding the crashed UFO. Nothing concerning any archaeological work on the Plains in 1947 showed up until Dick's telling letter to his advisor turned up in 2000.

*Figure 11.4. **Indian Artifacts**. A collection of Indian artifacts from various places. These were on display at the Eagle Guest Ranch restaurant in the late '90s. There are many private collections in the Plains area, but archaeologists need to find them at a site for scientific accuracy. Barney Barnett gave 5,000 or so projectile points, such as these that he collected during the Dust Bowl years in Oklahoma and Texas, to the Smithsonian.*

Why all this archaeological activity in the years 1947 to 1950? There was a flurry of activity in many areas of the west after World War II. Due to the distances that needed to be traveled, gas and tire rationing during World War II kept many anthropologists and archaeologists near their home universities. For example, the distance between Albuquerque and Reserve in 1947 was 210 miles, with half of it on unpaved roads. One individual who remembers those years, especially during the war, said that one could usually come up with gas for a medium-length trip, but tires of the day were subject to blowouts, especially on gravel roads, and they were very difficult to replace.

Another reason for the amount of archaeological activity in 1947 was the time required for funding. For a major dig, this could require a preliminary trip to take samplings of the archaeological layers, known as a stratographic survey. These results had to be analyzed before a major funding package could be put together and approved. Many of the area's digs were planned in 1946 or perhaps during World War II, but because of war delays and shortages, were not begun until 1947. Permits were also needed. In almost all cases researched, a permit to excavate was obtained from the U.S. Department of Agriculture, sometimes the U.S. Forest Service and, of course, private landowners when necessary.

From Martha Dick and material at the Smithsonian [15], we learned that Jeff Morris, Martha Dick's brother, worked with Herbert Dick on the Plains in later July for about a week prior to the Chaco Conference, and then they worked again after the Chaco Conference in August until the rains came in mid-September. We were able to locate Martha Dick in 2002. I had heard she was not in good health, so I declined to make contact with her at that time. I did not want what Martha Dick might say to be lost forever so I decided to give her phone number and address to Linda Moulton Howe, a UFO researcher, about 2007 when Linda moved from Pennsylvania to Albuquerque. Upon learning that she had not been contacted by 2011, I gave her name and address to Stanton Friedman. We also found that Jeff, her brother, who was also at the 1947 stratographic survey, was alive and living in Houston. I also gave his name and phone number to Stanton Friedman.

Yes, there were archaeologists on and around the Plains in the summer of 1947. Dr. Brew and Edward Danson no doubt on New Mexico Highway 12 to get to some of their destinations. Hurt and McKnight visited there when they could, but were both full-time students in 1946 and 1947. Both have "alibis" — McKnight was running a frozen custard shop in Albuquerque (similar to today's Dairy Queen) and Hurt was working in Alabama until about the middle of August. Students going to and from the Pine Lawn dig at Reserve were no doubt on Highway 12 that summer but none fit Barney Barnett's description of "a party of four or five archaeologists arriving together from the opposite direction." The Dick party, we know, were across the Plains at Bat Cave on the night of July 1 and 2, and would have arrived from the southeast, the direction of the cave.

## SURVEY RESULTS AT BAT CAVE

In his book, Dick said that, after the 1947 Bat Cave stratographic survey,[16] "I decided that the area would be worth exploring." Several universities and one foundation thought the project was promising, too. For the 1948 dig, the UNM contributed invaluable resources, including trucks and a student labor force.[17] There were about sixteen people in the 1948 party. At that time, Bat Cave and the surrounding area was still owned by the Hubbell Sheep and Cattle Company. The Bat Cave 1947 survey and expanded digs in 1948 and 1950 are American archaeological classics. I obtained Dick's book about Bat Cave through an inter-library loan, and was fascinated by it. In the 1948 dig, which lasted about two months, his party found many items used by early man.

Plant and animal remains at the cave enabled Dick and his team to determine some of the foodstuffs consumed by the early inhabitants. There were remains of 19 plant families and 39 genera.[18] The inhabitants of the Bat Cave area ate many kinds of seeds, roots, and berries, as well as corn, squash, and later, beans once pottery technology was developed. Some plants were edible; others, such as grass, were used for bedding, and yucca fibers for sandals and some garments. One of the most dramatic discoveries was pod corn or maize, which was thought at the time to be some 5,200 years old. (Later radio-carbon tests have reduced this time somewhat.) I was able to speak with Randall Montgomery who had been one of the UNM summer students to accompany Dick on his 1948 Bat Cave dig. Montgomery said he was among those who discovered the corn cobs.

Of the 30 or so types of mammals known to exist on the Plains, the bones and teeth of 13 different kinds were found in the cave.[19] They include the remains of bison, deer, antelope, and beaver, as well as the lowly wood rat and ground squirrel. Some charred remains were also found, carbon-dated to about 5,000 years old. The early cultural material at Bat Cave is associated with the Chiricahua stage of the Cochise culture.

There were 4,459 items found at the site, including over 1,300 projectile points, choppers, knives, scrapers, and drills; over 200 pottery fragments; and 69 bone tools. Also found were braided cordage, 14 baskets, whole or in part, and a buffalo hide bag containing beans. There were also 766 shelled corn cobs and 36 sandals, whole or in part. (Several sandals were very small, evidently for children.) There was also one adult-sized specimen, oversized, with caked mud on the bottom, apparently for walking in or around the soft and muddy edges of the playas.

In 1926, an African-American cowboy named George McJunkin found flint projectile points among extinct bison bones. The new category of Paleo-Indian was called Folsom man, named for the nearby town of Folsom in the northeast corner of New Mexico. Dick had hoped to find evidence that Folsom man had inhabited Bat Cave.

As glaciers melted during the Ice Age (approximately 11,000 years ago), water flowed into the San Augustin Basin. For over 4,000 years, the wave action of the lake eroded a very coarse agglomerate rock in the cliff face of the southeastern side of the lake basin. Evidence of Folsom man has been found in the lake bed to the north, but the cave was not formed as yet while the Folsom people were in the area. When they were there, they hunted the game-rich grassy areas around the north end of the receding lake, but no evidence of a permanent camp has been found.

Dick did find that Bat Cave had been occupied intermittently by other Paleo-Indians as far back as 5,200 years ago. After the Bat Cave dig, Dick was somewhat disappointed with the results. He happened to mention the corn cobs to Earl Smith, who was analyzing the botanical material found during the expedition. Many kinds of plant remains were found, including the cobs. A professor from Harvard, Paul C. Mangelsdorf, who was an expert in maize, recognized the maize find for what it was. Prior to Dick's Bat Cave dig, corn in North America was thought to be no older than 2,000 years. When the analysis was in concerning the carbon-dated corn cobs, charcoal, and wood fragments, Dick's find pushed this date back by 3,000 years, an extremely important archaeological discovery. This set off an entirely new field of studying North American Indian agriculture. Dick and his find became famous in articles, newspapers, and professional journals throughout the world in the late 1940s and early 1950s. Articles about Dick's find appeared for years in newspapers, and are still occasionally written.

In a previous chapter, Barnett related that he and the archaeologists were ordered to leave the crash site when the Army arrived. From another reliable source, we understand that the archaeologists' car (a pre-war Ford V-8) was parked about one-fourth of a mile from the site to the southeast). The Army tried to shoo everyone to the northwest towards the main highway. Apparently there was some conflict with the archaeologists (who were more than likely the Dick party) because they wanted to go back the way they had come towards the vehicle. We know the military attitude was very hostile towards civilians who got to the crash site ahead of them, and we believe they were quite intimidating.

We have reason to believe through Beth Danley (Fleck Danley's wife) that the intimidating and authoritative attitude of the Army was of enough concern that Fleck contacted his superiors at the USDA in Albuquerque. There are indications that Barney and Fleck went back to near the crash site to do "some surveying."

*Figure 11.5. **Socorro backyard stump** of a local geologist and Indian artifact collector, is decorated with metates and manos, food processing stones. Stones such as these are usually found at permanent campsites. There are also shoes for oxen, mules and horses, as well as other memorabilia found in the Plains area.*

This was quite within the realm of the Soil Conservation District's responsibilities. It is believed that the survey was to determine ownership of the section of land where the crash occurred. The author's research indicates that the land in 1947 and today belongs to the state of New Mexico and was leased as grazing land by a rancher living nearby. It is interesting to note that curiosity about the crash and bodies was very strong. We know that the Dick party of archaeologists came back to the Plains until the rains came in early September. The author discovered photographic archives (in 2000) at the Smithsonian Institution which show August 1947 photos of two members of the Dick party within a mile of the crash site from a high vantage point, with a clear view of the arroyo crash area. Martha Dick identified one of them as her brother Jeff. We assume the photographer was Herbert Dick. We know from his writings that he was doing research there at that time. With the Plains being 59 miles long and up to 15 miles wide, it stretches the law of chance considerably to have two parties who supposedly didn't know each other appearing within a mile of the crash site in a one-month period.

## UFO INTRIGUE

The illusion that Dick and his party had created when they all denied being at Bat Cave in early July had a rather dramatic effect on the future course and direction of UFO investigations. When Barney's sighting report was first published in *The Roswell Incident* in 1980, it was given a great deal of credibility by most researchers. A decade-long search for the archaeologists Barney had seen at the crash site turned up some time and energy-wasting blind alleys that eventually led to bitter disagreements among the researchers. The disagreements primarily occurred after a phone call to an *Unsolved Mysteries* program in 1990 produced a man in southwest, Missouri, who said he was at the crash site in 1947. This turns out to be the Gerald Anderson that Chuck Wade brought to the site in 2011. The author believes Anderson was at the crash site but differs on some of his recalled details. I have chosen to go mainly with Barnett's adult recollections but find some concordance with many of Anderson's recollections from the crash site. His description of the craft debris in the lower arroyo was outstanding. The foil shards we found buried (see Chapter 4) we think deals with the material Anderson described as "an explosion in an aluminum factory." Although introduced in 1947, post-war commercial aluminum foil did not make its appearance in west-central New Mexico until 1955.

Friedman and his researchers backed the man who would have been five at the time. Kevin Randall and his researchers took the opposite view that the young boy was remembering too much. The rather bitter feud was taken to the UFO community in papers and talks at UFO conferences and became rather wearing and divisive for UFO researchers. An organization called the Fund for UFO Research called all of the warring UFO investigators together for several conferences in Chicago in 1991 and 1992. The resulting 87-page report was published in 1992.[20] As a result of the friction between the Friedman and Randall camps on Anderson's story, neither side found the long-sought archaeologists. A wedge began to form between the originally-accepted Barney Barnett story and the Randall Roswell researchers.

During the late 80's and early 90's the Randle team of UFO researchers started moving Barnett around like a chess man. Although Gerald Anderson's story about the Plains crash occurred in the early 1990s. It wasn't too long before some mainline researchers found what they considered flaws and inconsistencies, and started discounting the story. We cannot evaluate various documents Gerald was said to have submitted to investigators because they are not available. There is still some animosity sensed by the author 20 years later from both sides of this conflict that originated in the early 1990's. I talked to several of those who had gone through this conflict. One man said, "We thought Anderson was probably telling the truth about the crash and bodies discovery." He said, "Some of us could and wanted to believe the crash story, which was plausible, but the conflict he got into with the die-hard skeptics turned his supporters away."

We located a paper published in the UFO community in 1991 known as *The Magdalena Fact Book* with the article titled "Why We Moved Barney Barnett from the Plains of San Agustin." The paper was authored by Kevin Randall and Don Schmitt. The Randall supporting researchers/investigators had a field day for several years. Some embraced the theory of Barnett's discovery near Roswell. And the natural assumption was to associate the archaeologists with a supposed UFO and bodies found by Barnett north of Roswell. A good example of this was when a well-known writer, Jim Marris, suggested in his 1997 book, *Alien Agenda* that, "Grady L. Barnett claimed he, along with some archaeologists who happened to be working north of Roswell, discovered wreckage and reported it." This certainly killed two birds with one stone: not only did the writers take Barnett 200 miles out of the Solado district, but he must have picked up the archaeologists we know were working in early July at Bat Cave, and taken them with him.

The discovery the author made in 2000 of Dick's party being at the Bat Cave site July 1 to July 14 may have made a difference prior to the friction between investigators, but was too late to change polarized opinions, many of which had gone into several publications and books. It wasn't so much Barnett's story that the researchers wanted to work into the Roswell scenario but his discovery of bodies, which made much more interesting copy for the researchers and ultimately the readers.

We cannot be absolutely positive that the archaeologists reported by Barney Barnett were the Dick party, but the evidence strongly suggests it. So does a statement to the author in an April 3, 1998 letter from Alice Knight, saying, "All I remember is his (Barney's) encountering the military people in a truck and meeting some folks on a "dig." This points to the Dick party who, as far as we know, were the only recorded archaeologists excavating on the Plains that summer. The term 'dig' is very specific. Alice did not say archaeologists looking for a site to excavate or that some people were out looking for Paleo-Indian sites. A dig, to any New Mexico person (especially Barney with his knowledge of Indian artifacts) means one or more professionally trained people conducting an excavation for purposes of discovering more about native Americans who had previously inhabited the area. The same definition also fits "a team doing archaeological research," as Barney told Vern Maltais.

Herbert Dick passed away in the early 1990s, in a state he loved and gave so much to. In the final analysis, we should ask ourselves, who among us with a college background or advanced degree would embarrass a prestigious university or jeopardize a chance to acquire a Ph.D. then or now? In retrospect, if the Dick party was involved, it was a very wise decision not to make public a flying saucer story that ultimately the government would deny or cover up, and few would believe.

# CHAPTER 11
# REFERENCES

1. Hurt, Wesley R. and McKnight, Daniel; "Archaeology of the San Augustin Plains: A Preliminary Report"; *American Antiquity;* 14(3); 1949; pp.172-94.

2. Ibid.

3. Dick, Herbert W.; Letter to J.O. Brew; Dec. 14, 1947; Acc #48-44, Upper Gila Reconnaissance; Records, Peabody Museum, Cambridge Mass.; p. 2.

4. Dick, Herbert W., *Bat Cave*, The School of American Research, monograph No. 2, Santa Fe, NM, p. vii.

5. *Socorro Chieftain;* Aug. 21, 1947; Socorro Publishing Co.; Hugh Thompson, publisher; Socorro, New Mexico; p. 4.

6. Brew J. Otis 19; "1947 Reconnaissance and the Proposed Upper Gila Expedition of the Peabody Museum of Harvard University"; *Elpalacio*, Vol 55, No. 7; 1948.

7. Ibid.

8. Brew; Op cit.

9. Dick Letter; Op cit.

10. Powers, William E.; "Basin and Shoreline Features of the Extinct Lake San Augustin, New Mexico"; *Journal of Geomorphology*, Vol. II, No. 4; Columbia University, New York, 1939; pp. 345-357.

11. Hurt, Wesley R. and McKnight, Daniel; Op cit pp 173-174.

12. Ibid.; p. 175.

13. Ibid.; p. 173.

14. Diary, Ruth Barnett; Entry August 20 and 21, 1947.

15. National Anthropological Archives, Smithsonian Institution, 900 Jefferson Drive, Washing DC.

16. Ibid., p. vii

17. Ibid., p. 88

18. Ibid., p. 90

19. Loc cit.

20. *Plains of San Augustin Controversy, July 1947: Gerald Anderson, Barney Barnett, and the Archaeologists.* J. Allen Hynek Center for UFO Studies, Chicago, Illinois, Fund for UFO Research, Washington, D.C. 1992

# CHAPTER 12

## ARTIFACT & CONTENTS — CLOSER EXAM

It took several months to begin the analysis on the artifact, as I will call the large HDPE piece. At first I was uncertain that it was not of animal origin. When I returned home, one of the first to see it was a veterinarian. He looked it over and said, "It is not any part of a small or large animal I have ever seen." He suggested that I take it to a veterinary department at a state college, but I decided not to pursue this avenue. Eventually, I took it to a college biology professor the next time I went through the university town. He examined it and felt it may not be from any biological source. Several weeks later, at another institution, several scientists looked at it and agreed that it seemed to have been subjected to a high heat source or an explosion.

## FIRST TESTS

One of the professors suggested I get a combustion analysis, which would compare the carbon, hydrogen, and nitrogen content. Since no offer to run any tests came from the university, I decided to work with commercial laboratories. This turned out to be the best way to get technical information. It costs an arm and a leg, but these laboratories not only give professional reports, but are much more precise and scientific. I would accept all the free advice I could get. This gave me a basis and a direction to investigate. I then took or sent the samples to commercial laboratories for analysis.

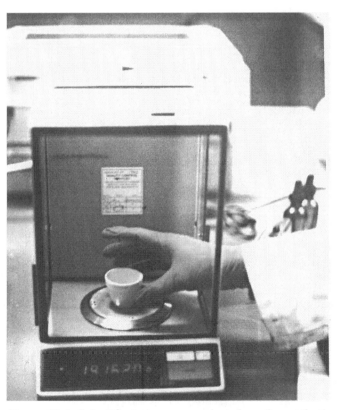

*Figure 12.1. **Scientific tests** were conducted on the artifact's exterior. This test determined purity of artifact material, using an analytical balance device.*

The first analysis was an ash test of one piece of an external layer. The laboratory then submitted two samples of the artifact to another laboratory for additional testing. The report is as follows:

Project No. EAGX-96-1229
Report No. 08-96-0418

**Dear Mr. Campbell:**

**Re:** Analysis performed on one (1) sample submitted on June 13, 1996 pursuant to your request

**Background:**

**A sample of unknown material was received for analysis to identify the composition of the material. The sample was composed of two distinct layers: a brown outer layer and a cream colored inner portion. We were requested to identify and compare the chemical composition of the two layers.**

**Each layer of the sample was analyzed by infrared spectroscopy in the micro-transmission mode. Identification of chemical constituents was made by comparison to reference spectra.**

**Conclusion:**

**The sample contained 0.47% ash. This indicates the sample is 99.53% organic or mostly composed of carbon based matter.**

**Although they differ in color, the brown and cream colored layers *are identical* in chemical composition. The unknown sample is basically *amorphous polyethylene*. The amorphous character of the polyethylene would indicate that the polymer has undergone a transition - to a more *random* state (perhaps by heat or chemical treatment). A trace of metal silicates were also found which would explain the small amount of ash.**

**Lead Chemist
Supervisor, Chemistry**[1]

The words amorphous polyethylene (above in italics) might be significant here. Very basic research on the Internet (Wikipedia) indicates that one form of amorphous polyethylene at temperatures above the glass transition level (130°C – 137°C) becomes amorphous polyethylene terephthalate. The fibers have been shown to possess optical and mechanical properties comparable with those of rubber.[2] The HDPE artifact apparently functioned, when it was operational, in a rubber-like mode. It certainly could not receive, dispense, or pump fluids of any kind in its present stiff, hardened state. The Internet information went on to indicate that amorphous materials are often prepared by rapid cooling of molten material such as glass, candle wax, and even cotton candy.[3] We will turn the research over to experts in this field, and hopefully it may help medical science some day with artificial organ transplants.

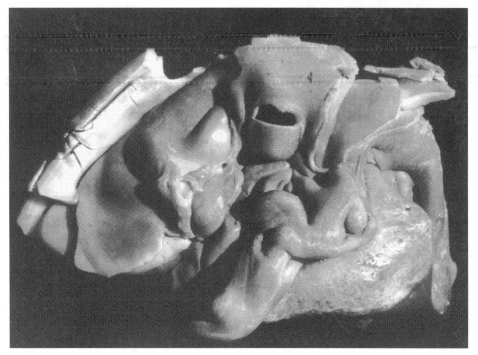

*Figure 12.2. The top side of artifact as it was found:* the edges were buried in the arroyo sand. Medical experts surmise it was a container of some sort, and may have also been a pump. The layers are made of an amorphous high density polyethylene (HDPE) without a crystalline structure. Originally thought to be soft, it is believed to have hardened or crosslinked in the sandy soil.

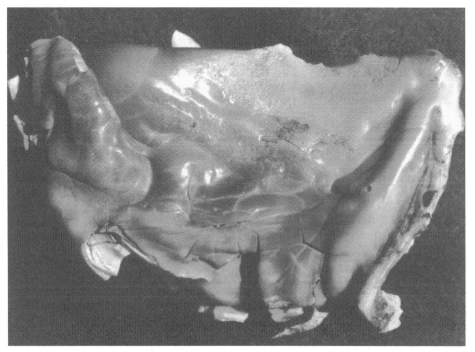

*Figure 12.3. The bottom of the artifact* is in good condition except it is chipped and broken on the top edge, (see Figure 12.9). Local livestock activity may account for this. Laboratory analysis indicates it had been in high heat or an explosion. Black specks on the lower right edge were analyzed by FT-IR and turn out to be congealed sulfonated natural and synthetic oil. This side was buried when found. Dimensions are 4 1/2 inches (11.5 cm) long by 3 inches (7.5 cm) wide, and weighed 4 ounces (113 grams).

## IN THE LAB

When the materials were tested by the two laboratories, as indicated by the letter from the lead chemist, the following tests were done:

1. Ash Test

A piece of the artifact was placed in a crucible and weighed on a 4 place analytical balance (a very sensitive scale) (Figure 12.1). The crucible was then preheated and placed in a muffle furnace at a temperature of 600°C or 1,100°F. The artifact material was then burned away, leaving the ash and any contaminants (dirt, sand, etc.) in the crucible. The crucible was cooled in a desiccator and weighed in the analytical balance again. The weight of the ash left in the crucible was 0.47 percent. This was subtracted from 100 percent of the original weight, leaving an organic material of 99.53 percent purity.

2. An FT-IR (Fourier Transmission - Infrared Spectroscopy) Test

This test was done in a second laboratory using a Nicholet Magna spectroscopy device with Omnic software. This test sent infrared light waves to bounce off the artifact pieces from the two separate layers. The spectrometer produced peak and valley graphs or a spectra of the two layers. The instrument looks at the resulting spectra of infrared colors of the object. These colors are determined by the vibration of the chemical bonds between the various elements in the compound, e.g., carbon, hydrogen, oxygen and nitrogen. The resulting spectra of colors is a fingerprint of a molecule. Every organic molecule has a distinct and individual spectrum (fingerprint). These spectra can be readily stored in a computer database for future comparison. The database in this computer contained at least 65,000 compounds. The search for each spectra takes about forty-five seconds. The search confirmed the preliminary finding of the ash test and revealed the spectrum found was the fingerprint of high density polyethylene, or more commonly known in the plastics industry as HDPE.[4]

A chemical engineering professor noted while looking at the wavelength graph that there were other bonds showing that were not polyethylene. He said, "Some of this is perhaps water or there may be some other material in the plastic. It may not be pure polyethylene, but one that has been packed, loaded, or filled with something else." Scientists who were able to spend some time with the artifact were quite interested in it. One offered these observations while inspecting the bottom side of the artifact. "The smooth material probably solidified in the air. It probably did not have a hot surface, but was in a hot environment of some sort, which softened the plastic and formed it in the shape it is in now. It looks like it came to rest on a textured surface, which took the imprint of it, while it hardened. As it cooled, it may have shrunk and thickened."

He thought it was definitely man-made and observed it has inside and outside layers. He thought it had been some kind of a container, noting, "It has a top and a neck, and it may have functioned on its side (because the bottom was a thicker material). It has hollow cavities and we can see the material inside folded and resting on itself." I told him that I had sent in some material for analysis that appeared to be a separate bonding layer between the outer and inner layers. This also turned out to be polyethylene.

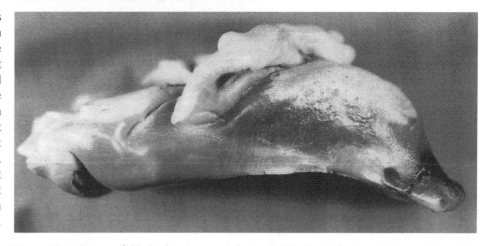

*Figure 12.4. **Bottom folded edge** intact, nicknamed "the duckie" after photos and samples were submitted to a national UFO group, and the author was told not to worry. This was only a "ceramic duck!" When found, golden brown bottom section was buried, and lighter area was above the arroyo sand. The darker area (on extreme right) indicates the polymer had started a melting process before cooling.*

## OBSERVATIONS – MORE TESTS

When having the metal foil shards analyzed, the technology of their manufacture was apparently far ahead of anything we had or needed. Were the manufacturers 100, 400, or 1,000 years ahead of known earth technology? I realized that if the artifact was a collapsed artificial body organ from an ET that, although it was made of polyethylene, it might have properties or manufacturing processes that could advance our medical science. One of the next efforts in this research is to get an analysis done to see if we have HDPE that was soft and flexible. Another question is whether this material would function inside an ET or human body without complications of rejection, which was the problem with many artificial body organs in our past. In other words, it may test as being HDPE, but was anything done to improve it as compared to how HDPE was made and functions today? Did this organ (if it is one) hold, receive, dispense, or pump any kind of body fluids? If it did, then would a similar one (provided we could manufacture it) be valuable to our medical science today? Did it have properties or one or more functions that today's HDPE possesses? If so, it could be very valuable to medical science.

While inspecting the artifact with a magnifying glass, I detected what I thought to be a solid rod, which may have run inside the artifact from the top of the duct or neck, down about two inches into an interior cavity. It seemed to be tapered from about two millimeters at the top to about three millimeters as far down as I could see it (Figure 12.13). It bent over 180 degrees when the artifact was apparently subjected to a heat source and collapsed in on itself. If it was a rod, I believe it may have acted as a stiffener to hold up or open the neck for what I thought might be a "drool" from a processing machine. All others who actually saw and handled it recognized that it was definitely a made object. "A bag of some sort," one doctor surmised. Those that handled it were forensic experts, college professors, chemists, and many associated with the medical profession. We proceeded to have it x-rayed at the advice of some professionals. One orthopedic surgeon who viewed the x-ray said, "If that is polyethylene, it had to have been impregnated with barium or calcium." He gave me an x-ray of a polyethylene and metal hip joint. The pin in the upper part of the femur in the hip joint was stainless steel with a polyethylene socket. What he pointed out was a synthetic glue surrounding the pin as it was attached to the femur. It had been impregnated with barium, which gave it a degree of opacity under the x-ray. This was a tantalizing point, but I was unable to tell if the artifact HDPE had anything added to it of this nature or if anything had been added to the HDPE or the material between the artifact layers.

I made an appointment with the head radiologist at a large medical center. He looked the artifact over and put the x-ray on a light box. Several physicians and technicians passed by and stopped. None had any idea what it could be. He did suggest that an MRI (Magnetic Resonance Imaging) might show something. He called a friend, who told him if I could get over to his office before closing, he would run the test. When I arrived, after a short wait, the man took me back into the room with the magnetron. Before I entered the room, he had me take my coins, car keys, etc., out of my pockets. He placed the artifact in the carriage of a Siemens MRI Magnetron, Model 42. He set the instrument at 1.0 TESLA magnet, and we retired to an adjoining room to watch two television monitors. It was a beautiful snow storm, but no image of any kind appeared. Apparently, to bounce a signal, the MRI needed hydrogen. Polyethylene is basically two-thirds hydrogen. Had the hydrogen undergone some change so it would show no image on the monitor? Was the amorphous question raised by the chemist a factor here?

# UFO CRASH AT SAN AUGUSTIN

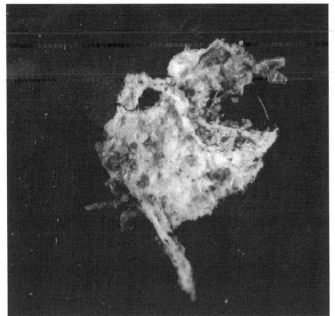

*Figure 12.5. **Piece of material** taken from artifact turned out to be cellulose starch. The starch was no larger than a pinhead. (Animal face is a coincidence.) Magnification x 100.*

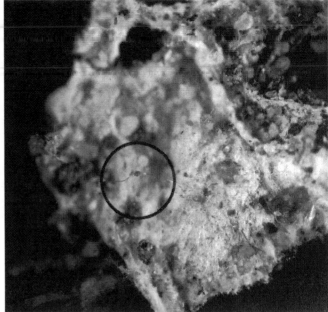

*Figure 12.7. **A gold wire** appeared upon additional magnification of the starch (see black circle.) Small round objects are microscopic grains of sand. Magnification x 8.*

*Figure 12.6. **Gold wire magnified**, again x 10, turns out to be flat and widened at top and bottom ends. A component of some type may be in the center. Several gold and copper wires were found on the piece of starch. These may have been residue from some type of mostly Bio-degradable monitoring system used in an alien body.*

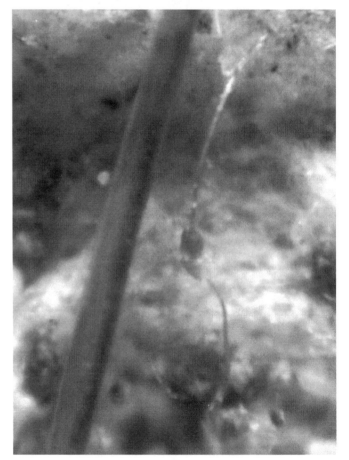

*Figure 12.8. **Gold wire and human hair**. Hair is about 70 microns wide. Gold wire is about 7 microns wide. Magnification x 10. At one time the artifact, whatever its function may have been created inside with starch to catch anomalies in whatever fluid may have passed through it.*

A medical librarian at a university sent me a recent medical device register. There were nearly a hundred companies making medical devices from LDPE and HDPE. The products included: sutures, prostheses, artificial skin, various types of membranes, heart valves, replacement blood vessels, artificial joints, contact lenses, and dentures. All we could deduce from these medical products was that polyethylene was a suitable material for long-range external and internal medical uses. This, of course, was not news to the medical people, but the sheer variety and versatility of the material to a layman was very educational.

There were three identified substances in and on the artifact that were not HDPE. The first one we will consider is the material that was embedded into the upper and mid-part of the right edge (Figure 12.3). The lab report is as follows:

**CONCLUSION: The gray-colored flakes are composed of a blend of sulfonated oils, both natural and synthetic. They might have originated from (or have been generated by) an incomplete combustion process, or possibly a residue from a larger source of oils. No inorganic compounds were found.**

## BODY FLUID AND STARCH

I asked the analytical chemist who did the work, if any of the oils could be specifically identified. He indicated that far more sample would be required to make that determination. I asked him to comment on the incomplete combustion process. He said, "Complete combustion in an uncontrolled environment is unlikely. The by-products were somehow sulfonated."[5] Here was a mystery that I could not pursue or refine. It seemed like a dead end.

I decided to go through my UFO material and search for any mention of sulfur. William S. Steinman, in his *UFO Crash at Aztec 1986*, tells of a UFO crash some thirty miles southwest of Laredo, Texas, in the Mexican state of Neuvo Leon. He relates that a naval photographer and four colleagues were flown to the UFO crash site in July of 1948. The photographers took some 500 pictures of what was reported to be an alien body and crashed craft. There was apparently no blood, but in what appeared to be veins of the alien, there was a light green fluid with a strong sulfur smell.

Leonard H. Stringfield, in his 1977 book *Situation Red, The UFO Siege,* tells of the investigation of an area in 1973 where a UFO was seen on the ground. A farmer fired some shots at the craft with a .30-06 rifle and the UFO took off. A state trooper was dispatched to the site and found an area about 150 feet in diameter glowing white. Later, about 1:30 a.m., several men arrived at the site and found that farm animals would go nowhere near where the UFO had been on the ground. Several of the men began to feel ill and one had trouble breathing. A strong smell of sulfur was in the air. The men left hastily.

Stringfield also related the story of a Dr. Epigoni, a retired physicist, who had worked on the Manhattan Project. Dr. Epigoni was working at Edwards Air Force Base in the late 1960s

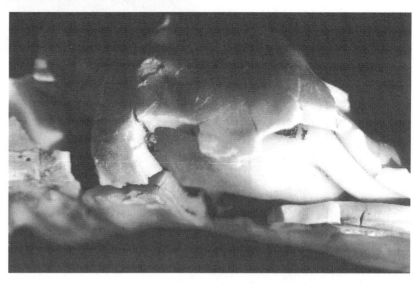

*Figure 12.9. **Enlargement** of broken edge reveals the starch material inside (left center). This tiny piece of starch held a gold wire. An enlargement of starch with additional wires is shown on the next page.*

and early 1970s, in charge of a program called "Blue Heaven." In his interview in 1990, he was asked about alien bodies and a craft held at Edwards AFB, California. The good doctor said they drew about one cubic centimeter of fluid from a body, put it under a microscope, and it looked like oil. Dr. Epigoni told the interviewer that he had been involved in briefing President Nixon and other top officials in the Oval Office in Washington, D.C., about the craft and alien bodies.

Here we have presented several reports about the odor of sulfur and one concerning oil. Is there any connection between the sulfonated natural and synthetic oil found on the outside of the artifact and these stories? It is anybody's guess. One of the forensic experts who helped with this project thought "this oil may have been embedded into the artifact during an explosion. He felt that "the surface of the artifact was softened by the heat," allowing the oil to be absorbed by the plastic. Upon cooling, this plastic/oil mixture took on its present hardened condition. There are signs that the artifact surface had reached the melting point at one time as the brown outer layer had started to run."[6]

The second material found associated with the artifact was starch. The starch appeared in two places, with more probably inside. The first concentration was inside the only opening I knew of, which is called the duct (Figure 12.11). Another piece of starch was visible just inside the broken area on the top backside of the artifact (Figure 12.9 above). The analytical chemist report was as follows:

**CONCLUSION**
**Sample # 5    Identification**
**Comments (Chemical Structure)**
**Starch and sand    Also $KI/I_2$ positive for starch**[7]

Confirmation for the presence of starch was made by using the $KI/I_2$ spot test. I asked the chemist to describe the spot test mentioned above. He said the test used a $KI/I_2$ solution, a solution

*Figure 12.10.* **Starch and copper wires** *2-3 microns wide. Starch was placed in a scanning electron microscope on an aluminum stub Light round areas are microscopic grains of quartz sand. Small square-ended rods may be starch fibers. Magnification x 100.*

of potassium tri-iodide which is amber in color. A drop of this liquid placed on starch would turn the liquid from amber to blue.

Research into starch turned up some interesting things. Starch is a rather basic building block and is manufactured during photosynthesis by plants. It is made by green plant leaves as a reserve food supply. Starch is stored in granules and is present in such organs as: the tubers of potatoes, the stem of the sago palm leaf, the seeds of corn, wheat, and rice, and the roots of the tapioca plant. Most commercial starch is made from corn and is used as a thickening agent in confections and baked goods. It is used in the brewing industry, in the manufacture of paper and cardboard, and in the textile industry.

Starch is also used in the polymer industries which make polyethylene. However, since the starch residue was found adhering to the inside of the artifact, we can only surmise that it was associated with whatever the artifact might have held when it undoubtedly was some form of container. Whether the starch served as a thickener for something or was a nutritional residue from something else, we just do not know.

In the research, I also learned that starch is insoluble in cold water. Apparently, weather, rain, snow and wind had little effect on the starch in the artifact. The duct side had been exposed for some time and the starch on the bottom side of the artifact was somewhat protected being buried in a half inch or so of sand. The artifact is translucent when held up to a strong light. One can look into the broken side for an inch or so before the view is blocked by folds. We are able to see more flaky material adhering to the plastic which looks like more starch, but is yet to be analyzed.

### THE GOLD WIRE AND THE REDHEADED CO-ED

One day I had the opportunity to view some of our small pieces of material under a strong microscope. We took some scrapings of the oil, previously mentioned, the wax pieces, some foil and what was left of the starch after it had been analyzed, a piece about two millimeters long and one millimeter wide. The compound microscope was an Olympus SZH 10 research microscope, which allowed us not only to see the material some forty-five times its actual size, but to photograph it with an attached camera. While studying the starch, I found a brilliant blue fiber on one edge. I decided to switch the lens to a stronger 2x magnification. A glint of something caught my eye. When I finally got the focus, a gold wire jumped out at me with something in the center.

The Olympus microscope I was using was in a local college biology lab, where I was working near several students. I chose a girl standing nearby who had red hair and asked her if she could help me. I explained that I needed a red hair to show up in my photo (Figure 12.8). After her initial surprise and some humorous conversation, she readily agreed. She even held one piece of her hair across the microscope slide while I photographed it.

When I looked at the photos, I decided to try and determine the size of the wire by comparing it to a hair. Hair is measured by a micron scale. Human head hair ranges from 30 to 120 microns

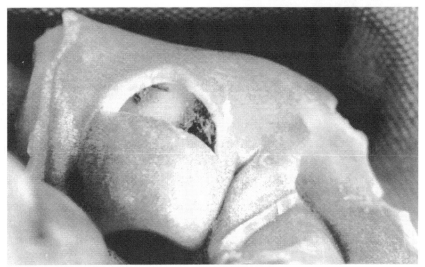

*Figure 12.11. **Magnified view** clearly shows the starchy substance inside. The duct as we call it was double lined, as was much of the upper part of the artifact. Jagged edges right and left indicate the artifact was torn from similar material.*

(0.03 to 0.120 millimeters). I assumed that the hair I used was somewhere in the mid (or 70) micron range. It appeared that it would take the width of ten gold wires to equal the hair, so my wire was about 7 microns wide. The wide piece in the center of the wire I judged to be about 28 microns wide. The length of the wire was some 700 microns or about 0.75 of a millimeter long. I could not be absolutely sure the wire was gold. Part of my background as an arts and crafts teacher was in jewelry making. I am familiar with gold and how it looks, although my work with the metal was limited. I cannot think of any other metal that would keep its luster like this in the elements, other than gold. Visual observations showed the wire to be flat and thin and wider by about fifty percent at one end. A scientist who was present that day, it was discovered, thought it was gold, and it looked like gold to another who viewed it later. Both scientists thought it was definitely man-made (some gold wire does occur in nature). One person knowledgeable in electronics thought that the wire piece in the center might be an electronic component connecting two wires. It has a slightly different look from different angles.

The Pentagon Insider Col. Philip Corso had access to alien autopsy reports from Walter Reed Hospital. The secret document that Corso referred to said that the alien brain was laced with IC's (integrated circuits). Might the microscopic wires found on the starch be from one of these integrated circuits? The gold wire was flat, which would indicate that it might have been on a tiny integrated circuit board and worked its way down into the artifact via body fluids.

Research into gold wire connected with UFO debris turned up on one occasion that is known to this writer. The case involved a UFO crash in Russia on a high hill overlooking Dal' Negorsky, U.S.S.R. in 1987. In Stringfield's *Status Report VI*, he related that investigators at the site determined that a temperature of 4,000°C to 25,000°C had been reached during the crash, and vegetation was affected by this radiation. Fragments of a microscopic, Alpha-Quartz netting was found. The thickness of each quartz thread was 17 microns (about one-third the diameter of a typical human hair). Inside each quartz thread was a gold wire. Stringfield went on to say that such technology and precision is just impossible for our present technological capability or our civilization. He continued, "Quartz is an ideal insulator and gold is an excellent conductor." Technology, as we know, constantly changes. When this research was being compiled, several scientists indicated that gold wires of 1 micron in width had been developed in the last few years and are used in some electronic components in the computer industry.

A year or so after having the starch identified, I was able to take it and some other items to a scanning electron microscope. The device used was a Zeiss Model 960 with an x-ray link. The starch readout can be seen below, labeled **Overall Scan Starch and Metals**. A forensic scientist gave his opinion of what the readout contained. He explained the x-ray fluorescence chart, activated under the scanning electron microscope, produces information about gradients of various metals. He said the dead time mentioned on all the SEM readouts was due to the percent of time required by the energy detector to recover from data collection. He went on to say that the peaks on the starch readout graph indicate the fluorescent spectra of several elements, including a high percentage of silicon, then potassium, iron, calcium, copper, and titanium with traces of sulfur and chlorine. He next turned his attention to an analysis of the copper wires in

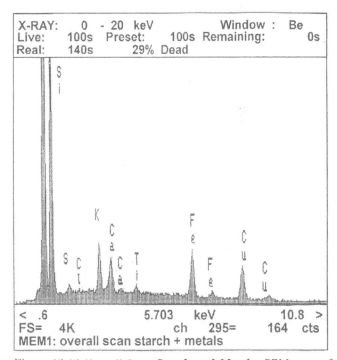

*Figure 12.12 **Overall Scan Starch and Metals:** SEM scan of starch from HDPE artifact. Embedded in starch was a cloth fiber with gold and copper wire fragments. Readout indicates that eight separate elements were present in the starch: Sulfur (Su), Chlorine (Cl), Potassium (K), Calcium (Ca), Titanium (Ti), Iron (Fe), Copper (Cu) (wire) and Silicon (Si). Some of the above elements were also in the soil. Unmarked high peak thought to be oxygen.*

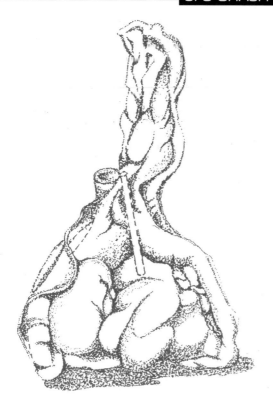

*Figure 12.13. **An Alien heart?** Called a vascular assist device by a medical school, how the above object may have looked before collapsing in on itself. Tapered rod in center may have served as a stiffener to keep the duct upright. Bio-medical experts think that the above artifact may have been the central part of an alien heart-lung combination.*

the starch. It was of different sizes, which appeared to be twisted and 1 to 2 microns wide in size (Figure 12.7). The readout of the first wire, which was flattened, indicated that it was embedded into the starch with a presence of copper and some chlorine and sulfur. The forensic scientist thought the chlorine and sulfur were not on the wire, but in the starch, which shows up in the background.

The scan for the second copper wire appeared to be a slightly different, smaller configuration, and showed the same as the overall starch scan, but a weaker signal. The scientist determined this by the absence of titanium and iron in the scan. Just after we enlarged one end of the second wire to 348 times its actual size and photographed it with a Polaroid camera, the wire disappeared from the scope. The operator said if any organic material nearby does not have the right coating, it could cause the item to "vaporize or flash off." Other items to be analyzed that day had a gold dust coating, but because we were analyzing metal wires, it was decided not to use the coating on the starch.

Although we found the above-mentioned additional smaller copper wires in the starch, the gold wire had disappeared. Our forensic man explained that it could have been lost in handling the starch from the storage vial to the aluminum stub. He said that microscopic metal tends to jump around and be attracted to other metals, such as handling tweezers or any metal nearby. We went back and looked at the storage vial and the lid on the aluminum stub box and still no gold wire. The wire was microscopic and could not be seen without magnification equipment. Our expert said it might have gotten into the SEM and "flashed off" before it could be recorded. It made me feel better when he said that was not uncommon and happened in laboratories all the time, but it was still quite a loss.

## A VAD & TUPPERWARE

We hope someday to find the resources and expertise to go into the artifact with a tiny endoscope. We believe from what we have seen, there is certainly more inside. We are not sure its contents will enlighten as to its function. We are sure the HDPE artifact, when it was functioning, was soft enough to work, but it also had to be stiff enough to stand or hang by itself. One scientist thought that the thickness on what we call the back side of the artifact indicated that whatever its function was, it may have occasionally been in a horizontal position. This would make sense if the HDPE artifact was in an ET and possibly in a reclining posture. What would be its age if it was part of an ET body? Certainly over 60 years since the Plains of San Augustin crash. Would it be 70, 80, or perhaps a 100 years old?

We believe it was made like an apple tart, starting out in a circle, folded in half, with its amenities added. See Figure 12.2. The edges were somehow affixed and slightly rolled. Once we find out what the main substance inside was, we will know more. Certainly more scientific work is needed. In 2008, the author was invited by the History Channel, *UFO Hunters* TV program to revisit the crash site. While on camera, the artifact, now housed in a protective plastic box, was given to the show personnel to be analyzed.

Scene two: the artifact duly arrives at the Seal Forensic Laboratories in Los Angeles, California, and is given a reasonably thorough analysis.

Conclusion: it was made of high density polyethylene HDPE. This is the exact diagnosis another lab had reached 11 years earlier. It was also determined that it had been in intense heat or an explosion, another factor established earlier. Its makeup was similar to, but apparently was made years before, the commercial product we know today as Tupperware.

Because of its contents, we can speculate that it was in some system or other, and foreign materials (the gold wire, copper wire, blue fiber, etc.) traveled into it.

Could it have been in an alien circulatory system? Possibly. Or was it some other kind of unknown body organ? According to the cattle inspectors who first saw the artifact, it was obviously was not from any local livestock. Considering where it was found, we could probably associate it with the other extraterrestrial finds.

We had the artifact evaluated at a large medical center in the southeast. They concluded that it might be a "vascular assist device" or VAD. In our medicine today, the VAD is also called a ventricle assist device. It is a mechanical pump that is implanted to help a weakened heart muscle to distribute blood throughout the body. Apparently its main use is to assist a heart patient who is waiting for a transplant. In an alien culture, perhaps many, many years ahead of us, it may have evolved to become a main pump for distribution of body fluids in alien clones. We hope future research will bring forward more data on this fascinating find.

# CHAPTER 12
## REFERENCES

1  Campbell, Art; Correspondence with laboratories concerning artifact analysis.

2  Wikipedia

3  Ibid.

4  Correspondence; Op cit.

5  Ibid.

6  Ibid.

7  Ibid.

The following references were valuable to this research:

*Chemistry, the Central Science*; Prentice Hall; Theodore Brown and Eugene H. La May.

*Thomas Registry of American Manufacturers*; 88th Edition; 1998; Thomas Publishing Co., New York, NY.

*American Plastic, A Cultural History*; Jefferey Meikle; University Press, New Brunswick, NJ.

"The Corso Files, Part II"; William A. Kent; *UFO Magazine*, Volume 7, No. 4.; August-September, 2002.

"Retrievals of the Third Kind. A Case Study of Alleged UFOs and Occupants in Military Custody"; Leonard H. Stringfield; MUFON Symposium, 1978.

# CHAPTER 13

## THE SHOE SOLE, WAX PIECES & WAFER

The most perplexing of any of the finds were the shoe soles. It was one thing to pick up foil and wax pieces, but quite another to gather a personal item that had actually been worn and used. I still have difficulty accepting that this shoe sole may be unique. Consequently, much more laboratory time and research have been put into this item. One would think a shoe sole would be easy to research; it was not. I eventually had to divide the research into four directions: general foot anatomy, infant foot growth and corresponding arch development, shoe manufacture and basic construction, and laboratory analysis of the shoe material. Questionnaires were sent to over thirty shoe companies regarding shoe manufacturing and materials. Shoe sales people, pediatricians, podiatrists and other professional orthopedic people were also contacted. I also acquired some material distributed by the National Shoe Retailers Association on foot anatomy and shoe fitting.

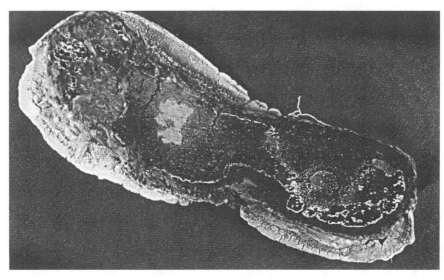

*Figure 13.1. **Small shoe** sole is extremely narrow. It measures 4 7/8 inches by 1 1/8 inches at the narrowest (12.6 centimeters x 2.8 centimeters). The foot that fit this shoe would have to be a 9A. No shoes are made narrower than 4A. A shoe store owner in Socorro stated, "That wouldn't fit any foot that ever walked in here."*

As was mentioned earlier in the book, a small shoe sole was found as I was going over the arroyo site one evening with a Geiger counter. I did not see it at first. On a rocky slope about ten yards away, I found the small dried rubber pieces referred to in a previous chapter. There were just a few slivers, which I placed in a plastic bag. A little while later I picked up the tiny sole, below the slope where I had found the above-mentioned material. It appeared similar in texture and color to the slivers, but due to gathering darkness I did not make the match until that night in my motel.

Subsequent tests indicate that the dried rubber slivers and the bottom of the small sole were of identical material. Again, water enters our picture. The small sole was found in an area next to the present arroyo water course. When the arroyo braid plugged up with sand and silt, the water was diverted down and alongside a narrow track used by cattle grazing in the arroyo. As I mentioned before, I was there one day during an afternoon rain. The water diverted down this track and stood at a depth of five or six inches, as it spread out into a delta-like configuration, covering the spot where the small shoe sole was found the year before. Whether the sole was washed in over the years or was uncovered where it lay, I do not know. The slope above the sole was rocky with light vegetation. How the tiny pieces of sole ended up ten yards up the slope is a mystery.

The layers of the sole were caked with dried dirt, which I carefully washed out through a handkerchief at the motel. Some fibers and threads came out, as well as several tiny tulip-like seed affairs about 1/16-inch long or one half centimeter. These, on later investigation, turned out to be tumbleweed seed.

The scientific investigation to identify the sole layers was extensive. Three laboratories were involved; two on the west coast and one on the east. Diagnostic procedures on the nine layers of the small shoe sole included infrared spectroscopy, differential scanning calorimetry, polarized light microscopy and the use of a scanning electron microscope. The equipment was necessary to analyze six different materials in the multi-layered sole. I really do not know where the shoe sole was made, but if it was made here, it would have to be after the 1940s when polyurethane foam was developed. This foam rubber in shoes did not come into general use until the 1960s, according to shoe manufacturers contacted. The nylon thread which held the straps together, again, was not available until the 1940s or so. Nylon (first made in 1935) was just beginning to be used before World War II. The most popular product was ladies nylon stockings. Nylon production was quickly changed to parachutes for the military. An analytical chemist from the Portland, Oregon area said, "**The white pigment in the bottom rubber layer was identified using polarized light microscopy and energy dispersive x-ray spectroscopy.**" These tests revealed several materials including the presence of titanium dioxide, a white pigment first discovered in 1916, but not available until 1920.

The first laboratory to look at the sole was a forensic laboratory where a technician separated the nine layers and mounted them on glass slides. The lab revealed that **six of the nine layers were made from cellulose plant fiber. The third laboratory analyzed the exact material in each layer, identifying six of the layers as a sort of cellulose/linen/cotton, four of woven cloth and two layers of compressed cellulose fibers embedded in other materials.** The bottom of the small sole became a study in itself. The first laboratory report follows:

## LAB RESULTS

Richard E. Bisbing of McCrone Associates said:

"**The sample submitted was brittle and a somewhat plastic gray material found on the bottom of the small sole.**

The material was identified as plant "pitch" and/or oils(like lecithin), and a minor amount of cellulose and Kaolin-type clay(possibly of the plant "pitch" to be some type of rubber). We were unable to confirm this since the material is not homogeneous." [1]

I asked Mr. Bisbing for a description of the PLM process. The laboratory scientist sent with his report a description of how polarized light microscopy helped identify the small outsole, as

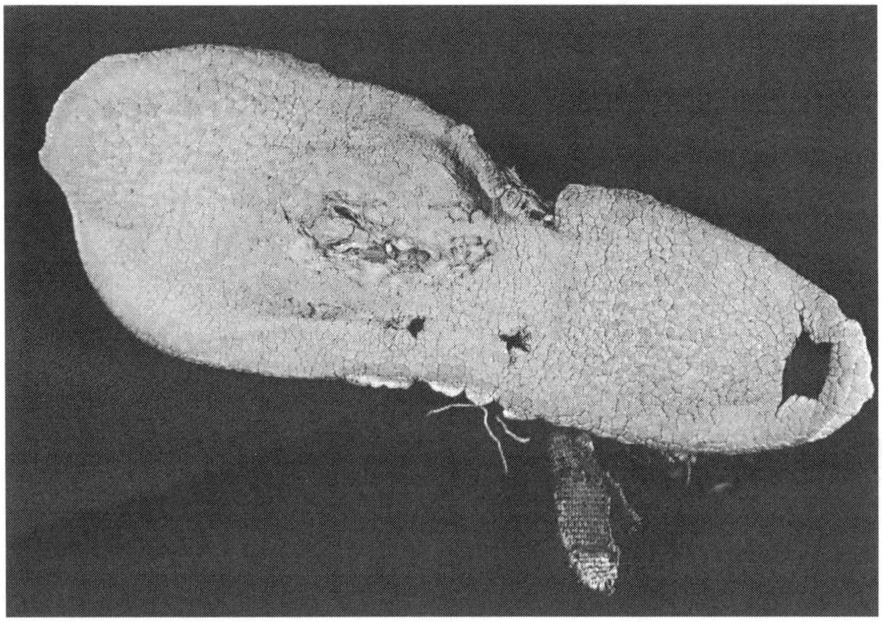

*Figure 13.2. **Small shoe bottom** has an interesting crackle surface that tends to grip when walked on. Shoe is very worn in middle. Note cotton fibers coming through the rubber on the shoe bottom. Cloth fibers were discovered above the rubber layer and other material. Cloth fragments were between layers and are believed to be the outer cover and liner of the shoe.*

follows: "PLM particles are identified by observing transparency or opacity, texture (internal and surface) and color, by transmitted and reflected light." He went on to say that "coupled with experience, PLM generally enables the identification of more than 95% of the materials in a given sample." [2]

This data came from the laboratory that was involved in the Shroud of Turin research, mentioned earlier. The same scientist at McCrone Laboratories was also intrigued by the bottom layer of the small sole and found it "most interesting." He reports:

"It was relatively hard and brittle and appears to me to be mostly filler. It can best be described as a degraded filled material. I also analyzed the sample using an electron microscope with energy dispersive x-ray spectroscopy in order to determine the elements present. The tan material contains a clay, calcium (carbonate), titanium dioxide and a lead pigment."

Sizing the small foot template to fit the shoe, turned out to be a fascinating experience. This time a children's Bannock foot sizing chart was used. Again the foot experts were stumped. Over a dozen podiatrists were consulted in this shoe project. One podiatrist wrote;

"The length is 4 7/8 inches or an infant size 4 1/2. The width is impossible, which comes to a 9A. This is a non existent size. We don't even have a 1A size for a foot of this width, let alone a 9A. We could not give any further size details, because we've never had a need to measure anything this small." [3]

I found it interesting that over 75 percent of the small sole layers were made of plant fibers, and the bottom sole contained a variety of materials including plant pitch. My first thought was that the back-to-nature and natural fiber people will love this.

For the little sole, I constructed a questionnaire which I sent to shoe companies that have a wide variety of footwear, including casual and dress, and moccasins to cowboy boots. I asked if the rubber formulation above was used today. This question was answered in various ways depending upon the type of footwear manufactured. In general, the answer was yes, natural rubber materials are used today in shoe manufacture, but they are generally giving way to more sophisticated petro-chemicals. In a process the manufacturers called soling, one company listed six different petro-chemical rubber-like products used to make outsoles.

## SHOE COMPANIES

Is it possible that the little shoe had been larger and then shrank to its present dimensions? I posed this question to a forensic scientist

*Figure 13.3. **Cloth fibers** (also shown in Figure 13.2 above) correspond to layers 3 and 4 on the artist's drawing of the shoe sole (next page). The top layer is believed to have been the shoe liner that was sewn to a strap (right) to make the cotton shoe layers stiffer. The bottom piece, of a different weave, was apparently the outside of the shoe, and was a faded red or maroon in color.*

## Artist's Drawing of Shoe Sole

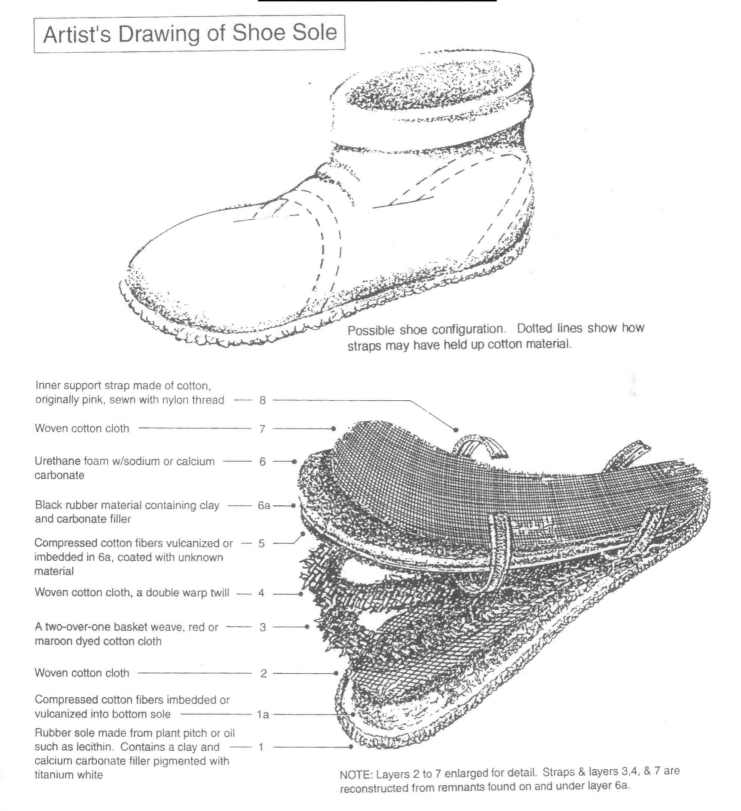

Possible shoe configuration. Dotted lines show how straps may have held up cotton material.

| Layer | Description |
|---|---|
| 8 | Inner support strap made of cotton, originally pink, sewn with nylon thread |
| 7 | Woven cotton cloth |
| 6 | Urethane foam w/sodium or calcium carbonate |
| 6a | Black rubber material containing clay and carbonate filler |
| 5 | Compressed cotton fibers vulcanized or imbedded in 6a, coated with unknown material |
| 4 | Woven cotton cloth, a double warp twill |
| 3 | A two-over-one basket weave, red or maroon dyed cotton cloth |
| 2 | Woven cotton cloth |
| 1a | Compressed cotton fibers imbedded or vulcanized into bottom sole |
| 1 | Rubber sole made from plant pitch or oil such as lecithin. Contains a clay and calcium carbonate filler pigmented with titanium white |

NOTE: Layers 2 to 7 enlarged for detail. Straps & layers 3, 4, & 7 are reconstructed from remnants found on and under layer 6a.

*Figure 13.4. **Artist's drawing** of small shoe layers and possible configuration showing how shoe may have looked. The numbered layers represent nine different levels of the sole, as identified by FT-IR, PLM and other scientific analyses. Apparently, the outside of the shoe consists of two layers of woven cotton cloth (3) and (4) above. It is also thought that the cloth straps formed a support for the exterior shoe shape. Of the nine layers, six are all or part cotton. Lab reports state that the sole is of a filled natural rubber from "plant pitch or oils like lecithin." **(Layer 6)** was a soft cushioning material of urethane form, covered with a finely woven cotton cloth.*

who first examined the sole. He replied that some of the cotton layers may have shrunk if they had not been firmly attached to the bottom rubber sole and the top urethane layer. He believed that although these materials had hardened and partially curled, they had essentially maintained their original volume, as well as had the cotton between them. He also pointed out that the nylon thread which was present in the strap remains, could not shrink.

Another question to the shoe manufacturers was, who might have manufactured the little sole? All who responded said they did not know. One said that only a manufacturer's logo could really tell. A consultant for a very large shoe manufacturer in New Jersey was quite familiar with shoe imports. He thought the item was extremely well made and suggested that it may have been "cobbled by hand" in Spain or the south of France. He thought it bore strong resemblance to "alpargates" (soft-soled shoes worn by Indians of Bolivia or Peru).

One shoe company representative wrote, "I am not familiar with the bone structure of *pygmies*; perhaps it might fit in that category." He went on to say, "The width of the white insole pattern has a forepart *too narrow to fit any foot*." (The shoe industry calls the narrow point in front of the heel "the forepart.") Many infants/toddlers grow through this 2.8 centimeter or 1 1/8 inch width on the forepart of their feet, and some larger infants have this width very early in their growth process. The area, however, widens considerably as the child matures and develops. A child under fifteen months usually requires a half-size footwear change in less than two months, although he or she may not be walking.

The sole was too complex in manufacture to have been an orthopedic shoe, and it had a narrowness in front of the heel totally out of proportion to its length. Professionals seemed baffled.

I contacted Dr. Roger Leir, who is a practicing podiatrist and UFO researcher from southern California. He wrote, "I have been in practice now for over 35 years, and have never seen a child's shoe of this dimension. Certainly a shoe of four inches is not uncommon, but a width of 1 1/2 inches *is impossible by human standards*." [4] Note: Dr. Leir measured his widths across the wider instep. Now here was a real mystery – a shoe sole of a narrowness that is less than any size manufactured, less than any size sold in shoe stores, and a size that was "impossible by human standards!" I spent almost two years trying to identify and classify the shoe sole. In the end I could identify how it was constructed, the size and color, the materials used, and even the wear pattern on the sole, but I have yet to discover what or who wore this tiny shoe sole that was of a length to fit a one- or two-year-old, but of a width no wider than an infant's foot. Other questions remained as well. When and by what means had this shoe arrived in the arroyo? With all the data I have collected, I still have to classify the little shoe sole as a complete unknown.

The inner sole top, which consists of the combined layers (excluding the lower rubber sole), measures 12.3 centimeters (4 7/8 inches) long and 3.8 centimeters (1 1/2 inches) wide at the ball of the foot. Several views of it can be seen in the photo pages. I constructed a sample pattern slightly smaller than the sole, which would have been close to the size foot that fit the shoe. I made a tracing and drawing of a two-year-old child's foot of the same length. I placed the Arroyo innersole template over the child's foot drawing. This, in turn, was placed over a child's Bannock foot measuring chart and the photo in Figure 13.5 was the result. The arroyo inner sole template, when placed on top of a modern child's shoe template, was about 35 percent narrower. The layer drawing, as shown in Figure 13.4, is a representation of the actual layers in the shoe sole. It is certain that layer 3 in the drawing was dyed red and was once the exterior cotton part, of an upper shoe. Layer 4 may have been an inner layer under the exterior.

One or two of the shoe companies that were contacted sent information about shoe manufacturing, which was helpful. I assumed from the questionnaire answers that the small sole was well made. From manufacturers' information, I learned that good shoes are made on "lasts" of the shoe that determine the overall shoe shape and fitting characteristics. A last, basically, is a foot-sized solid piece that remains in the shoe during its entire manufacture. Different length and width lasts give us the different shoe sizes. We could locate no children's shoe manufacturer who had lasts of this length and width.

Of the basic six types of shoe manufacturing, the little shoe seemed to be a combination of the "stuck-on" method used in everyday and casual footwear, and the "direct molded" process used in children's shoes, and for everyday general purpose, heavy duty, industrial, and water-resistant footwear. The investigation of the small sole shows that the bottom apparently curved up around the perimeter of the shoe for 1/4-inch or so, covering the other seven layers. There is also strong evidence that the bottom rubber sole was vulcanized (a process of heating rubber to improve its strength and elasticity). This would tend to have melded the rubber sole with the cotton fibers, as seen in layer 1 and 1a of (Figure 13.4). Cotton fibers in layer 5 are also embedded into layer 6a. There may have been some vulcanizing here or some type of heat process in the production of the 6a layer that could not be identified.

The weaving style of the layers indicates that all of the cloth layers (2, 3, 4, and 7), were standard one-ply weaves. Layers 2, 3 and 7 were basket weaves, and layer 4 was a double warp twill weave. It might be interesting to the reader to know that these weaves were also used on sandals of yucca fiber by the early inhabitants of the Plains of San Augustin some 4,000 years ago. Also, in Figure 13.3, note the remains of a double sewn cloth strap of some kind. There are indications that there were two. These were not straps as used on sandals today, but were placed at the wide part of the foot and near the heel to support the soft outer fabrics from the inside. Similar supports for the cloth fabric shoes can be found in modern footwear. The only thread on the sole not made of cotton is a two-ply nylon thread which sewed the folded-over strap to itself, and also attached it to the outer exterior layers of the shoe. The nylon thread was probably added for strength. How the cloth straps may have been integrated into the shoe construction appears in the mock-up drawing of what the shoe may have looked like.

Apparently, the outer shoe fabric had been dyed red. One observer thought that the color of the outside shoe layer may have been maroon. The laboratory that analyzed the fibers said that when teased apart, the outside of the yarn appears uniformly red, but it did appear that the inner fibers had not retained as much dye as the outer fibers. Close examination of the inner support straps found alternating threads of red and white, which would have originally appeared pink.

## UFO LITERATURE

This finding touched off another search into the UFO literature to see what colors, if any, might be associated with extra-terrestrials or ETs. My main source was the Leonard H. Stringfield Status Reports. Stringfield's work and findings are considered about as objective as written UFO material can be. In his *Status Report III*, written in 1982, he lists several colors of alleged ET uniforms. From a MUFON report of 1980, one of the ETs appeared to be wearing a tight-fitting dull metallic green uniform.[5] A report from CUFOS in the 1980s, referred to a sighting in 1944 in which an alien was also wearing a tight-fitting green uniform. An Air Force man had an experience in 1954 in which he and some companions saw some aliens in tight-fitting blue uniforms. Stringfield reported in Status Report V, 1989, that an Air Force photographer was flown to Norton Air Force Base in California in 1973 to film three dead aliens dressed in blue uniforms similar to flight suits. Another of Stringfield's reports, written in 1978, included an abstract from a Mr. T., a civilian in a high technical position. Mr. T. had seen a film in 1953 showing aliens again wearing tight-fitting suits. This time the suits were in a pastel color. In Stringfield's *Situation Red*, published in 1977, he lists another film seen by a civilian at Fort Monmouth, New Jersey, in 1953. This time the clothing on the small bodies was yellow, one-piece uniforms. In Travis Walton's *Fire in the Sky*, 1997, he lists aliens in "velvety blue uniforms" and some in orange surgical gowns. Many reported sightings in recent years list aliens in gray uniforms with no shoes.

In *The Day After Roswell*, Corso suggests that the ET pilots themselves may be part of the electrical circuitry in a flight suit that was molecularly aligned. This is an intriguing thought, and it may have something to do with some of the reports of metallic fabric flight suits. Aluminum was a component of the gray material found at the site. When the metal foil shards were tested in 2010, each piece we found had from 39 to 56 elements in a coating on one side. All foil shards had different elements, possibly signifying a separate use. We wonder if these flight suits may have had special elements in the suit to help the alien crew in flying or communication efforts. It is doubtful, however, that rubber-soled shoes would be a part of this process. The bottom of the small shoe sole does have an unusual wear pattern though (Figure 13.2). The rubber seems to have been worn more in the middle section of the sole than at either end. This might suggest that it rested on something harder and that some force or weight had been applied, causing the worn spot; for instance, a foot pedal of some sort or repeated trips up and down a ladder, etc.

I have kept an eye out in the past several years for anything in the UFO literature pertaining to footwear. As one might expect, people's claims range from the ridiculous to the sublime. Aliens are reported to sport a wide variety of shoes and boots. These include calf-high rubber boots commonly seen in barnyards, Three Musketeer knee-highs, wide ring affairs as might be seen on the Michelin Man, as well as World War II-type combat boots and elf-like boots turned up at the toe! One of the more interesting reports of an alien footwear was given by Travis Walton. He observed (with the blue uniforms mentioned previously), slightly loose pants tucked into soft dull black boots with a moccasin-type sole. (Several shoe company representatives thought the small sole to be from a moccasin-type shoe.) Walton also described another type of foot gear of a pinkish-tan which is about as close as this research comes to red or maroon.

Some of the more serious reports describing footwear indicate something similar to the drawing found in the upper part of Figure 13.4. The shoe design shown has some basis in fact. If the two cloth layers of the shoe encompassed the whole foot and were joined at the top, it would make sense that they folded down to create the look as shown. The shoe sole might also have been the extreme bottom part of a garment. One of the first to examine the shoe sole was a forensic scientist whose laboratory identifies shoes and other crime scene evidence. His comments lend some credence to this theory of the sole possibly being part of a garment. He thinks the fabric scraps remaining between the layers may have been cut on the outside edges. After having the sole in his possession for several months, however, he stated, "If I were asked in a court of law to identify this shoe sole, I could not." He said he had seen lots of canvas and nylon shoes over the years, but commented, "Most of the cloth shoes I have seen were slippers, but you would not need nine sophisticated layers for a slipper!" He went on to say, "This shoe was well insulated, tough and extremely well-made."

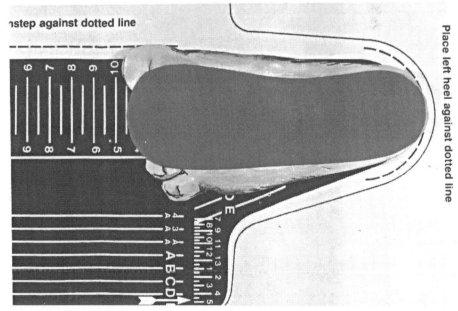

*Figure 13.5.* **This Bannock child's shoe** *measuring chart is shown with a template representing the found shoe sole superimposed over an infant/toddler foot representation. A human toddler's foot of the same length is five sizes wider than our found shoe sole.*

## HUMAN FOOT DEVELOPMENT

In order to obtain some background information, I decided to research human foot development. The foot turned out to be much more complex than it would at first seem to be.

The human foot has 26 bones, 33 joints, and over 100 ligaments. The foot has two main functions: (1) Supporting body weight, and (2) propelling the body forward with a levering action. The foot's flexibility allows it to adapt to uneven surfaces.

The development of the foot during gestation and into the toddler stage was interesting reading, but it was obvious that the foot that fit the small shoe sole was much narrower than anything described in the medical literature. The toddler's foot is somewhat more chubby and wider in proportion to that of older children, and appears flat until the child begins to walk and put weight on the foot. As the normal child develops and spends more time upright (from about one year to two-and-one-half years or so), the foot begins to lose excess fat and an arch develops from more usage. A good deal of cartilage, which is present at birth, starts hardening into bone as the child ages. It was interesting to read about the development of the arch, but there is no indication of an arch visible on the small or large shoe soles.

I sent for some information about the foot development of children of small stature. One study from Hong Kong dealt with the change of foot size with weight-bearing. This study of 2,829 children measured the linear foot growth of boys and girls from age three to eighteen. Another study of 586 children from Japan compared foot length with age and height. The results were consistent with other studies from the United States, showing that normal foot growth in length and width was in synchrony with the advent of walking and the development of the body as a whole. I had hoped to find something in the small foot information that might shed some light on the stature and foot growth of smaller than average children. Instead, I found just the opposite.

Many of the families in the Plains area are Hispanic, having been in New Mexico for generations. Gilbert Amijo (interviewed in Reserve), moved to the tiny town of Horse Springs when his parents bought a small store there in the 1940s. He said that most of the children he knew went barefoot most of the time and that their feet were so very tough, resulting in what he called "alligator" feet. He recalls getting his first pair of shoes in 1947 when he was ten. Gilbert said, "Some children did not own their own shoes until they were confirmed in the church or graduated from the eighth grade." Boys' first shoes were usually work boots, while girls often got Mary Jane-type shoes that are somewhat open on top. I asked another man who had grown up in the area about sport or canvas shoes. He had attended high school in Magdalena and later in Socorro. He replied that the first sport canvas type shoes he remembers were high-top black tennis shoes worn by basketball players. He did not recall anyone he knew wearing sport-type shoes of canvas or cotton until they went out for sports in junior high or high school. Apparently going barefoot when the body is still growing helps develop strong arches.

A Socorro man who had been selling shoes for over forty years said that because children in the area went barefoot a lot, their feet became wider and required wider shoes. He found that the narrow shoes did not sell and kept very few in stock. He was glad the "jogging type shoe" comes in standard widths today, which helps him keep better track of his inventory. These shoes have no half sizes. He reported that he did not sell as many shoes as he would have liked to Hispanic families when the children were young because they were usually a part of a larger family and had hand-me-downs available. When I showed him the small shoe sole, he turned it over in his hands and said, "That wouldn't fit any foot that ever walked in here."

When I sent out the questionnaires to the shoe companies, I also enclosed a photo of the small shoe sole, the composite drawing of the layers, and the foot template seen in Figure 13.5. When sizes are referred to in the shoe industry, import sizes must also be taken into consideration. One manufacturer sent a conversion chart for American, French and Italian sizes which differ from each other. I included a question about what age child might wear an insole the size of the small foot sole template. The answers varied. In general, the length would fit a one to two-year-old, but the narrow heel width, according to some professionals, was no wider than that of a young infant.

The urethane foam rubber of the sole's top layer is now hard and crumbly. I believe this layer was originally soft and spongy, but later hardened in the elements. Research indicated that polyurethanes are resins that produce solvent-resistant coatings and rubbers. This rubber can be made either soft or rigid. The soft material is used in a wide variety of materials including crash pads in cars, insulation, sound-proofing, rug-backing, shoe-padding, and other cushioning material.

One of the avenues I pursued was through certified pedorthists who do custom shoe modifications for medical purposes. I was able to learn the custom shoes of today use various gluing procedures to make or modify special orthopedic shoes. Stitching is sometimes used with leather and composition soles. Apparently, the shoe sole had not been involved in any orthopedic construction. I learned that children's shoes go through evolutions. For some time, the shoes were rigid and controlled. Today they are flexible with the least arch support available. One pedorthist who viewed the small sole said, "I have worked on children's and adult foot sizes for years and have never seen a size this small or narrow." He then said, "Some of the fibers in one or two of the layers were as fine as cobwebs."

None of the pedorthists felt the small sole was made for a deformed foot. They also do not think the sole was custom made. The small sole was a obviously a production model of some kind, judging from the lamination of one or two layers that required special machinery and materials. Although the top of the tiny sole had at one time been soft, the bottom sole material was probably harder for durable wear and traction. Finding the small shoe bottom to be worn very thin, was very interesting. One remarked, "It looks at least 30 to 40 years old, but has the characteristics and construction of a shoe of modern comfort." One manufacturer referred to the Chinese custom, before the turn of the 20th century, of binding women's feet to make them smaller. Although this changed the size dramatically, the Chinese foot was apparently short and stubby, but not narrow.

## ET BODIES

In interviews over the years, Vern Maltais described the alien bodies his good friend Barney Barnett had told him about. Barney said that he had stood very close to the aliens. On another occasion,

Barney said the bodies seemed intact under their clothing. Barney was describing the intact aliens laying outside the craft. He noted that a body or so was inside. They may have been mangled inside their uniforms, but there was no way to tell from his vantage point outside. The HDPE artifact had to have been ripped from an alien interior. Perhaps there was a body or so under the center floor of the craft out of sight. There were no stains on the uniforms indicating injuries or breaks in the skin. The heads had no visible wounds. Maltais had the feeling that it was painful for Barney to talk about. Before leaving the scene, Barney may have been asked to give his name and address. He later told witnesses that he had been contacted three times about keeping quiet, up to the mid 1960s. There are few credible descriptions of aliens, but to corroborate parts of Barney's description, a government manual known as SOM1-01 surfaced a few years ago. There is some conjecture in some UFO circles about its authenticity. However, Barney's description is similar to two ET types listed in this manual. One has a large head and a 3 1/2 to 4-foot body. The other has wide-set eyes.

The second source is Leonard Stringfield's *Status Report VI*, p. 127. His witness was a doctor who allegedly did some autopsy work on ETs. He lists the body at just over four feet tall with an enlarged pear-shaped head with eyes Mongoloid in appearance.[6] The same source gives some descriptions of the ET feet. The SOM1-01 manual lists one entity known as an EBE type II. The hands have three tapering digits and a long thumb; the feet have four toes that are joined together with a membrane. When Colonel Corso was stationed at Ft. Riley, Kansas, he said that the alien he saw had thin legs and feet. While I was doing medical research on foot growth, I learned of a condition known as syndactyly, a condition in which two or more toes are held together with a skin membrane. The condition is rarely a problem, and surgical separation of the toes is usually unnecessary. A slightly different description of something similar is given by Stringfield's

*Figure 13.7. **Fibrous tape** of wafer edge cut-away, reveals interior as a charcoal type material. The rather temporary makeup of this item suggests something that is disposable after some period of use. One scientist thought it may be a filter of some sort. Magnification x 5.*

doctor, mentioned earlier. He said that the foot he saw was covered with skin somewhat reminiscent of a sock. However, an x-ray showed modified bone structure. I was intrigued by another of Stringfield's sources who examined two humanoids over seven feet tall (*Status Report III*, p. 29). These were described as having very small toes and very flat feet.[7] The shoe sole I found in the arroyo, had no sign of arch supports.

## THE WAFER

Early in 1998, I was able to take several items to a scanning electron microscope. I had prepared two samples from this wafer, which was given to me by a ranch wife. One was a piece of outer ceramic-like hard-flaked material from the top side and the other sample was some of the gray fibrous material I was able to pull off the fibrous material that was wrapped in a narrow band around the outside of the wafer and appeared to be a separate material. Near one corner of the wafer on edge, I could see what appeared to be a loose corner of a layer of the fabric, reminiscent of a gauze-type tape protruding. I took this to be an end and extracted fiber samples from here. I also cut away a section of the fibrous material and noted that 99 percent of the wafer thickness was a charcoal-like material (See Figure 13.7). I had learned to expect the unexpected and I was not to be disappointed when the analysis came back. The ceramic-like flaky material was mounted on an aluminum stub and put into a SEM. Two terraces or levels of the flake were analyzed at a lab.

I had asked the operator to mount the outermost side down so we might pick up more of the interior material. This worked out well, as I believe the materials from terrace 2 were from deeper in the wafer. The next SEM analysis was performed on the gray, fiber-like material which, with the naked eye, we took to be some form of cloth or tape. It turned out to be short strands of compressed material containing a lot of zinc.

I made some simple observations but could go no further. The terrace #1, of the hard flaked material was probably

*Figure 13.6. **Wafer top**. Measuring 1 inch (2-1/2 cm) square and about 1/4 inch (1/2 cm) thick. Found at the site by a rancher whose wife gave it to author. From the top, it appears to be ceramic. Magnification x 2.*

*Figure 13.8. **Strange wax pieces** found near the pit. FT-IR analysis revealed that wax is of a vegetable type similar to carnauba wax. Magnification x 2. There is a strong possibility that the T-bar is the top 2/3 of a wax I-beam. Partial circle mark on the left end is believed to be a mold mark.*

I scanned the UFO materials for a description of anything similar. In the 1964 Socorro, New Mexico, UFO report, I read that some metal scrapings were taken from a rock and eventually wound up at NASA. They were found to be a zinc-iron alloy. This was obviously something made for a purpose. It resembled nothing anyone I had talked to could recognize. As far as familiarity goes, the materials and construction seem a little crude but not earthly. It must have had a purpose and a function but for now, the wafer will remain a mystery, classified in the unknown.

## THE WAX MATERIAL

The first piece of what turned out to be wax was found in the initial pit under the cross (See Chapter 3). I found more on the surface during the project, especially after a rain. I had three samples of wax analyzed: one from each of the three samples I had found, as well as the T-bar from the pit and several other pieces from the surface nearby. I was sure the one found about fifteen feet from where the artifact was discovered was a piece of the artifact. The color and texture was the same. I was totally surprised when the below report came back wax. Waaa...x? "Who would make anything out of wax?" my wife said. I walked around the house for several days mumbling waaa..x? The report on the wax pieces is as follows:

an outside top covering, composed of the oxygen, zinc, aluminum, and silicon. I had no idea what the zinc-like material might have done. It could have been insulation for what went on in the wafer or part of a conductor material for some other use. I later determined that the wafer had two flaky top layers with charcoal on the bottom. This may have been a filter of some sort. The zinc material seemed to be some kind of tape holding the two flaky layers to the charcoal.

### BACKGROUND

My correspondence and samples sent to the analytical chemist Utzman received this reply: "Five samples were received for identification. These materials differed in color and shape. All were quite small in size. No additional information was provided concerning their possible origin or sample environment."

### SUMMARY

The chemist indicated:

> "Samples 6, 8 & 7 are essentially identical in composition. All three are waxes, a *Carnauba* variety, which essentially means they are an exudate derived from some type of plant. In contrast, paraffin type waxes are derived from petroleum sources. These wax samples are definitely not

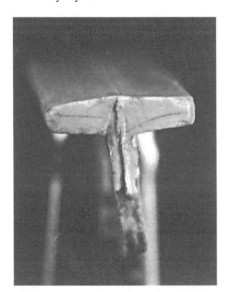

*Figure 13.9. **T-bar end** piece shows a protrusion which may have been attached to another piece. Magnification x 3.*

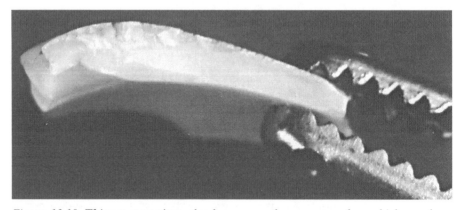

*Figure 13.10. **Thinner wax piece** also has a smooth concave surface which may have also curved around something. Magnification x 3.*

petroleum based. I am unable to be more specific as to the origin of these wax samples, since most waxes of this variety are very similar in their basic chemical structure.

"All samples were analyzed as received, in the condensed phase using FT-IR spectroscopy. Nondestructive analysis was also performed using FT-IR microscopy in the reflection and transmission mode. Identification of chemical components were made by comparison to reference infrared-spectra.

"The samples 6, 8 and 7, refer to the wax in the photograph [in Figure 13.8]. It appears that the waxes were a carnauba variety plant-based wax and they varied in saturation and strength. Webster's Collegiate Dictionary defines car-nau-ba as *'a fan-leafed palm of Brazil that has an edible root and yields a useful leaf fiber and carnauba wax.'* It goes on to define carnauba wax as *"a hard brittle high-melting wax from the leaves of the carnauba palm used chiefly in polishes."* Other literature searched said it had an "outstanding ability to gel higher concentrations of organic solvents and oils." This was desirable in paste polishes and carbon inks. Other uses include dental impression compounds, phonograph records, confections, candles, lipstick, floor wax emulsions and one of the most commonly seen uses, as a wax on apples."

This was a tough one; how does one determine the possible use of flakes of wax...? I could not. I searched through UFO material again and came up with several possible references to wax. Bob Lazar, who said he had worked for the government back-engineering a flying saucer propulsion system, had been allowed inside a craft. Bob observed there were no sharp corners; every device in the craft had a rounded corner. He observed it was almost as if everything had been finished with wax, then melted. I contacted Mr. Lazar and sent him photos of the wax pieces. I did not hear from him directly. However, a close friend of his said he had asked Lazar about his statement. Lazar said something to the effect that he did not say it was made of wax, it just looked like it. This posed an interesting question. If there was a coating of wax inside a UFO cabin, was it an insulator to keep the occupants out of direct contact with energized metal surfaces?

William S. Steinman, author of *UFO Crash at Aztec: A Well Kept Secret*, relates that an interview was given Dr. Berthold E. Schwarz, M.D., in his two-volume *UFO Dynamics* book. In Book II, Dr. Schwarz interviewed a former intelligence officer who said he had been inside a UFO cabin. He indicated the craft inside was dull silver. He said it was "not paint" and it had a "dull brown coating over it." The officer said if you hit it, you would get the impression there was something under it. Our wax pieces were light tan in color. Might they have been darker at one time, and over the years the arroyo had bleached out their original color?

I found one other interesting piece of information that might concern the use of wax in or around UFOs. Leonard H. Stringfield formerly worked with the Air Defense Command of the USAF, and served in a 5th Air Force intelligence unit in World War II. Mr. Stringfield wrote several publications concerning UFOs. He relates in his 1982 Status Report III, an interview of a highly placed military man who was present when the military tried to open a sealed crashed UFO. They eventually succeeded by poking a rod in the right place through a tiny hole in a porthole. Eventually a door opened. "It appeared in a place where there were no seams or other indications of a door." This door opening was almost as if the material of the craft had liquefied and then solidified again. Was wax somehow being used as a sealant?

Is there any connection between Bob Lazar's observation of the interior seemingly coated with slightly melted wax or the military officer interviewed by Dr. Schwarz who indicated the surface inside the cabin was not painted, but "had a dull brown coating over it"? Are these men talking about the same or similar coating inside two UFOs? How about the witness quoted in Stringfield's *Status Report III*, referring to a door opening where a seam seemed to liquefy, then solidify again? Could any or all of these observations somehow be connected to wax? I would be inclined to write all the above off as coincidence if it had not been for the T-bar which was made from the same kind of wax found a number of feet away from the other wax pieces and 20 centimeters under the soil. My best guess regarding the T-bar was that it was part of an extremely light structural device, not meant to be permanent, like something that would burn up in an aerial flair, etc.

Could the light "balsawood-like debris" reported at one or two other UFO sites have been wax? We will probably never know, but the three different pieces and densities of carnauba-like wax found in the Plains of San Augustin arroyo certainly add some dimension and questions to the mystery.

## REVIEW

In the end, I was able to match the small shoe sole with the length of a known foot, but not the corresponding width. Despite consulting knowledgeable individuals in the shoe industry, I was unable to locate any corresponding size or manufacturers' lists of that size used in shoe production. The small sole length was for a one to two-year-old infant. One podiatrist commented on the extremely narrow width of the small shoe sole. "The shoe width is off." It would not fit the bone structure of a foot this length.

The Roswell Crash and a few others where artifacts were found conditioned us to expect exotic materials. As far as we know, the Plains crash is one of the few crashes where interior craft materials were found. We also found in the arroyo remnants of several separate pieces of metal structure probably connected with the exterior and possible antigravity/propulsion system of the craft. What looked like discarded tin or aluminum foil at our site turned out to be coated foil-like material of varying thickness. Scientists believe the sophisticated coatings of elements on each foil shard could not have been created with 1947 technology.

Our HDPE body part, if a body part, may be the only one in civilian hands, and is definitely unique. Now we have a shoe sole with natural rubber and cotton. The surprise here is the cellulose cotton is not exotic but some of it appears to be coated with an unknown substance. We also found a natural rubber coating compound on the gray fabric scrap. Here and there in UFO literature, various odors (some chemical) have been reported from some witnesses near the "grays." This is not necessarily universal, but it is possible that the EBEs with some artificial body parts lack sweat glands such as ours. Cotton may have been found to be an ideal material to help counteract concerns in this area. Somehow it is reassuring to know that the cellulose product we know as cotton is produced elsewhere in the universe by the forces of nature we revere and respect.

## CHAPTER 13
### REFERENCES

1. Campbell, Art; Private correspondence from shoe companies; 1999.

2. Bisbing, Richard, E.; Report 21, May 1999; "Small Shoe Sole Analysis," McCrone Associates, Inc., p. 1

3. Campbell, Art; Private correspondence from podiatrists; 1998–2000.

4. Campbell, Art; Private correspondence from Dr. Leir; 1998–2001.

5. Stringfield, Leonard H.; "UFO Crash Retrievals: Amassing the Evidence"; Status Report III; Cincinnati, OH; 1982; p. 6.

6. Ibid.; VI; p. 127.

7. Ibid.; III; p. 29.

Some interesting reading in this area included:

*The Atlas of Human Anatomy* by Frank Netter, MD, CIBA-GEIGY Corporation; Summit, NJ, 1989

*Foot and Ankle Source Book* by David N. Tremaine, MD, and M. Elias, House Contemporary Books, Chicago, IL

*How Shoes are Made*, Strata Footwear Technology Center, North Hamptonshire, United Kingdom, 1997.

# CHAPTER 14

## THE IMPACT — OUT OF SIGHT, OUT OF MIND

We had always assumed that the bottom of the UFO was intact and it slid along in the arroyo at relatively low speed until it stopped on the knoll or small rise at the end of its skid. We also assumed that whatever we found in the 135-foot-long gap area had exited through the split Barnett had mentioned. But apparently this was not the case. One of Barnett's descriptions for the craft was a sort of "dirty stainless steel, relatively intact." In his description of the craft, he did not mention nor may not have remembered extensive damage to the craft's bottom.

The metal piece, shown as W-6 (see Figure 4.11), was tested in 2010. W-6 had a coating of some 43 elements and was a cast (mostly aluminum) piece. Chuck Wade led another crew back to the site in June 2011 and found more matching big pieces of metal and other smaller pieces that came to be known as "the Motherlode." The depth of these larger pieces varied from three inches to twelve inches. All were found with a metal detector, including the honeycomb, found the first morning. All in all it was a pretty good haul. When it first hit this upper part of the arroyo, large pieces of metal from several layers were probably ripped from the bottom of the craft, and may have been driven into the ground by the forward momentum.

The larger, very hard piece we found in 2004 was apparently from the exterior top of the craft. It should be said here that the pieces were in no way any kind of memory metal as reported so prominently at the Roswell crash site. We should not get memory metal confused with the extremely tough metal Marcel describes. These were obviously two different types of metal.

We believe the craft, when it impacted and lost some material from the bottom, skipped at least once. The initial impact might also have split open the craft, as reported by Barney Barnett. We do not know if the split was vertical or horizontal, but it is fair to say that the thinner metal (foil), shoe sole, and artifact (if body part) came from the interior of the craft. The second impact may have resulted in an explosion, scattering debris that the military missed and we found. We need to remember at this point that the arroyo, for the most part, was loose soil and sand. Any disturbance would have kicked up a big cloud of dust and dirt. When it settled, it may have covered smaller debris items. There is evidence we will discuss later that indicates some biological material was buried for a time nearby.

## THE GAP

It took a long time to make any sense out of the gap. Originally, I explored the idea that something had cut a swath through it, killing the buck brush. This thought lead to my use of a Geiger counter and the taking of the twenty-one soil samples. Finding that the soil was different in the center of the rectangle was a mystery. The suggestion by an agricultural consultant that the center portion of the gap might be fill was intriguing.

When the larger pieces of the UFO were found, they were in an area that lined up several hundred feet above the initial gap. For the large pieces of material to have eluded the military's initial inspection of the site, they had to have been underground, out of sight. Perhaps contour leveling equipment at the site had stirred up the UFO debris with human trash, which locals had dumped in the arroyo for years. It would have been impossible to distinguish this trash from the debris from the craft underground with the metal detectors available in those days (if they were used). We believe eventually a great deal of water, wind, and soil erosion took away the overburden, which began to uncover the shallower buried UFO debris, some of which we found. Today the soil in the area is only two to three inches deep, which makes recovery of the material easier.

When we started working in the softer ground of the lower gap in the mid 1990s, we had always assumed that the gap was a crossover between arroyo braids, but several factors did not fit into this scenario. A crossover occurs when one braid (a temporary channel) has a blockage and water is forced over to another adjoining braid. All other crossovers were narrow and could accommodate only a narrow meandering stream with irregular sides. The gap was five times as wide and had no signs of running water or an old channel. The buck brush edge, especially on one side, was in a straight line almost like a hedge.

We flew over the gap in 1997. Before the flight, we marked both sides of the gap in one-foot-wide aerial flagging tape. After examining the aerial photos, we noticed a hump in the middle indicating higher ground in the center, which rules out any flowing water as a cause of the gap. When we flew over the arroyo, the vegetation had returned years before, after the military had re-contoured the area. We finally had to concede that while filling in the lower arroyo crash trench, the buck brush roots sustained sufficient damage to prevent any future growth.

When I found the HDPE artifact, it was half buried. A weather line can be determined here and there, running horizontally around the perimeter. This would indicate that the artifact had been exposed to the elements for several years or more. It was found near some buck brush, which had slowed the surface erosion. I had the impression that at one time it had been totally covered. Other man-made debris in the gap area was also partially or totally exposed. There is enough slope in and around the gap for rain water that does not soak in to seep off, taking a little of the surface soil with it, uncovering material near the surface. This fact was borne out when I pulled out of the sandy soil a wad of metal foil still partially buried. One edge had appeared above the surface after a hard rain. I had been over the same spot thoroughly not an hour before and had seen nothing. If I bought into the fill theory that the center of the gap was made up of soil brought in from elsewhere in the arroyo, this would mean that the material I had found on the surface was in or perhaps under that fill. Chuck Wade and I were to find 14 or 15 pieces of these foil shards.

One of the complications of getting smaller pieces from the lower arroyo gap area was that they were underground and out of sight. The previously mentioned flash flood came down the arroyo some two weeks after the site's clean-up by the military. The flash flood water, as it crossed under the highway, was some four feet deep in a narrow, swift channel about twelve feet wide. The water washed out an approach to the highway culvert, and the road was closed for a day or so. By the time the water reached the crash site, it was much lower and had diverted itself into the current unblocked braid, an inch or so in depth. This braid had branched off the main channel and cut through the gap area where we were to find the valuable metal shards and some other material. The braid, no doubt, was covered over with silt, leaving the debris to erode out or was shallow enough for us to find it with the Navajo crew 50 to 60 years later.

The dynamics of the arroyo are as follows: as debris, sand, dirt, rocks, etc., move to lower ground, blockages occur, forcing water to back up and spill over the braid ditches, partially covering nearby low ground with standing water. This creates a delta-like effect over the low ground below the site, and much of the terrain remained in this condition for a day or so until the water evaporated or soaked into the ground. This could have very well caused some damage to the clean-up orchestrated by the Army earlier that month. This running and standing water combined with a hard rain, could have compacted or washed away some loose soil and exposed previously buried UFO crash debris. During a first or second clean-up, the gap soil may have been moved or altered. In doing so, it is possible that previously buried UFO crash debris was moved into the gap closer to the surface where it was found after erosion had exposed it, some years later. Cleaning up all the debris would have been impossible in the ever-changing arroyo. One ranch hand living nearby may have been asked to help pick up stray debris from what the Army termed a "missile crash." This will be alluded to in the next chapter.

After the craft was removed and the flash flood occurred, I doubt seriously if the Army went over the site a second time. More than likely, civilians and civilian equipment were used in a second clean-up, especially after the flash flood, to grade the soil back to the land's natural contours, if needed. Two ideal pieces of equipment to use in a broad flat area such as the arroyo would be a county road grader and a disc harrow, available at most ranches. One road grader was stationed in the area. Almost all secondary roads were gravel in those days. It is possible that the gap was bare of buck brush because most of the roots were taken out by the use of leveling equipment, apparently filling in the shallow trench. There is a bulldozer-cut road coming into the crash area, which may have accommodated a flat-bed truck to take the craft out.

It is not known if Barney Barnett had anything to do with the crash aftermath, but Ruth's diary shows him up in the "high country" on July 2nd, and also a few days after this date, and again several days after the flash flood. He would have been an ideal liaison between the Army and any civilians helping clean up the site. He knew about the crash, as he had been asked to cooperate in keeping the incident quiet. He knew about the roads and had worked with the property owners. He may have been the only one at the crash site who knew the lease holder of the land where the crash occurred. In his capacity as soil conservation engineer, he also had knowledge of where equipment was located and who was trustworthy and qualified to operate it. If any equipment was used, it may have been operated by civilians. It is doubtful, due to time constraints, the military brought in their own equipment with qualified operators. If this was the first post-war UFO discovery, it is very possible some of those involved in the clean-up and removal of the craft and bodies did not have high security clearances...that would not come until later.

One factor is very difficult to understand concerning the Barnett Plains of San Augustin crash scenario, and that is the military contacts with Barney over a 20-year period. We have a number of questions, but very few answers. Was the nature of these contacts partly geographical? Was it that Barney lived closer to the site than any of the other civilian witnesses? Did he have some knowledge of the flying saucer, its occupants, or anything that was not generally known that could threaten the national security at the time? Barney had access to the site, he knew the back roads, and he probably passed the site on the road periodically while he was still working before his retirement in 1959. There is evidence that the site had in-the-know visitors between 1947 and the 1960s or so. Did Barney's retirement precipitate renewed military interest in the site?

Were the military personnel concerned about the erosion vulnerability of the arroyo and the shifting channel braids? And one more important question arises. Was there someone in the west central New Mexico area who kept track of any of the original witnesses and/or occasionally visited and monitored the crash site? About the time Barney said he was being "harassed" a famous UFO sighting/landing took place about two miles from Barnett's house in Socorro. In April of 1964, police officer Lonnie Zamora, while chasing a teenage speeder, went up a gravel road west of Socorro. He found an egg-shaped UFO sitting in the sagebrush. The UFO took off, leaving Zamora quite shaken. We are sure Barney knew Zamora, but doubt if Zamora knew of Barney's experience. It is very interesting why Zamora was allowed to talk freely about his encounter while Barnett was admonished for speaking about his find.

We knew from experience at the site that any traumatic event could be blurred by clouds of sandy loam soil. This would be especially true after a UFO impact, skipping and skidding for several hundred feet. We believe the impact site may have been visited frequently at first for a few years to see if any evidence had been exposed by rains or winds. According to Colonel William Leed, Barney says he was interrogated several times in the 1960s. Would this interest in the discoverer of the site have extended to the site itself? We believe this is quite possible. After Barney's retirement, he was no longer a federal employee. Those monitoring him may have felt he might more readily tell others about the crash and the site.

I do not think all the pieces are in place for the above questions to be thoroughly answered. Did Barney have more responsibility and involvement beyond that of just being a witness? I find it hard to believe that a responsible man with progressive health problems, in his sixties and seventies, was contacted and interrogated simply because he told a few friends and relatives of his Plains experience over a 22-year period. We need to remember that at least ten to twelve adult civilians knew of this incident, plus any military men. We know five or six people in the archaeological party knew of the crash. I do not think the authorities monitored Barney so closely that they knew when or with whom he confided. Perhaps the other witnesses were so scattered that they could not easily be located. Also, Barney was close and convenient.

Over the years, as older military intelligence personnel who monitored the site were transferred, promoted, or retired, new men were assigned to take their place. The Plains crash event would interest anyone (it certainly did William Leed). And, as part of the background, the personnel may have wanted to visit the crash scene and any witnesses who lived nearby. When talking to Leed in 1967, Barney used the phrase "was interrogated." This does not sound like a particularly pleasant process, but perhaps, as Barney aged and his health declined, he may have felt more vulnerable. In his later years, these meetings may have seemed progressively more threatening. Indications are that the crash experience was a highly traumatic event to both Barney and Ruth.

**THE HDPE ARTIFACT**
This item was the very first item found. It took almost a year to determine how and where to have it analyzed. I originally thought that the artifact was a common material. It seemed that way when initially tested, but it turns out to be rather exotic in its origin, use, and function. When tested, it was found that the object had no crystalline structure, a fact which baffles scientists I spoke with. It was first speculated by the initial laboratory scientists that the crystalline structure may have been altered by heat or chemical reaction. This possibility has been checked out, and scientists agree that three sophisticated layers inside/outside, etc. might show differing degrees of crystalline structure, but the object would not be totally lacking in crystalline structure, as this object obviously is. Several said they thought it had been a soft container of some sort, but it would not function if it had a crystalline structure as we know it.

Many have come to associate HDPE with certain products, one of which is Tupperware. It is flexible, durable, and easily sanitized for foods. Another use of HDPE is in children's play equipment, from sliding boards to playhouses and wheeled vehicles. One college professor, knowledgeable in polymers, said, "Of what use would an HDPE item be, if it was not stiff?" I'm not sure of the answer, but I know that the HDPE artifact was soft, pliable, and easily expandable.

Some time ago, I was reading a book by Dr. Roger K. Leir, D.P. (podiatrist.)[1] Dr. Leir and a surgery team have been specializing in removing solid small objects from humans, which are thought to be alien implants of some sort. He mentions that the implants (some not much larger than cantaloupe seeds) do not have a crystalline structure, and the atoms are not aligned. Dr. Leir's implant removal research also involves some anomalies of amorphous magnetic iron metals. He did not think we had the knowledge to make metals of this type. Dr. Leir has hypothesized, I believe with some wisdom, that the body may not recognize non-crystalline items embedded under the skin and therefore does not reject these anomalous objects. If this is the case, then it would make sense to make artificial body organs of a non-crystalline material.

HDPE is also used for hundreds of medical purposes.

*Figure 14.1.* **Walter Reed Medical Center,** *(formerly Walter Reed Army Hospital) Washington, DC, as it appeared in the 1940s and 50s. New section in rear finished during the Viet Nam war. Col. Corso said that a secret lab here conducted medical research on alien bodies. Photo courtesy of Walter Reed Medical Center. Col. Corso photo* UFO Magazine.

It is highly resistant to body fluids and has uses within the body, which does not reject this material. It is not really a stretch of the imagination to recognize that we have developed an ideal material that, with some improvements and modifications, might be used by another society for similar purposes. It took several years to complete the testing of materials including the starch found inside as well as the flakes adhering to the outside, which were composed of congealed natural and synthetic sulfonated oil. It had originally been thought that this oil may have come from the inside of the artifact, but it is believed now that due to the trauma of high heat or explosion the oil may have come from another nearby material.

The starch found inside was very interesting. The configuration of the artifact suggests that it may have contained a liquid solution that may have moved in and out. If this was the case, the starch may have been a thickener in another liquid, or perhaps a suspending agent or emulsifier. Whether the starch thickened the material in the bag or was in some solution that entered the bag from another source is not known. In any case, over time, the bag was, no doubt, a repository of some kind, and it could obviously expand and contract, depending on its use.

One scientist who looked at the artifact was an eastern college professor with a background in chemical engineering. This gentleman noted that the bottom had been flat and was thicker and thought the artifact may have functioned on its side, rather than in a hanging position. This may fit in with the hypothesis that the aliens are reclining in their ships or that the ships are flying in such a configuration that their torsos are horizontal, similar to that of our astronauts while they are in take-off mode. My impression is that the object was at the terminus of some system, and various materials entered it. This thought is partly borne out by the discovery of the microscopic wires, a blue fiber and possibly other materials adhering to the starch. (See Figures 12.5–12.8) If the starch was in a solution to solidify or pick up foreign objects, it did its job well.

In the late 1990s, astronaut John Glenn, in his seventies, returned to space. New technology enabled him to swallow a capsule containing a thermometer and transmitter that relayed his body core temperature and other vital signs to a receiver on his belt. In this manner, important information about his body's responses was then transmitted to earth. Within the last few years science has been involved in the development of micro-technology and integrated circuits at a molecular level.

The concept of a monitoring device traveling through the body is not as unique as it seems. Around the turn of the millennium, a British gastroenterologist developed a camera in a capsule that may serve to replace painful colonoscopies and other uncomfortable diagnostic examinations. It seems wireless medical imaging and diagnostics are not far away. As a result of the research on tiny digital chips, a camera in capsule form has been developed tiny enough to be swallowed. It has no film or moving parts, and its size is 11mm x 16 mm. Inside is a lens, a flashing light-emitting diode (LED), tiny digital camera chip, batteries, a radio frequency transmitter, and an antenna. After being swallowed and activated, the light diode flashes, sending a photo, by radio waves, through the body to a sensor worn on a belt and connected to a computer. As the camera works its way through the digestive system, it can record any abnormalities found there.

## COLONEL CORSO

In a nutshell, Colonel Philip J. Corso was a career Army officer who served in Italy during WWII and the intelligence staff of General Douglas McArthur in the Korean War. In the early 1960s he was given a unique assignment at the Pentagon. Assigned to foreign technology in the Army research and development program at the Pentagon, he was put in charge of the Roswell files. This included some technical equipment that he said was given to private industry for military and civilian development. In essence he became a whistleblower on the military secrecy of the Roswell UFO coverup scenarios. Some of Colonel Corso's credentials and hard-to-believe story were initially challenged by UFO researchers. Documents obtained via the Freedom of Information Act confirmed his background and most of his story. He co-authored the best-selling book *The Day After Roswell* with William Birns. Colonel Corso became respected by UFO researchers and much admired by many. He died in 1998.

In Corso's book, he mentions the crash on the San Augustin Plains. Readers may recall his previous position at the Pentagon, and as advisor to the National Security Council and President Eisenhower. Corso gave several interviews. In his book and interviews, he connects the Roswell event with the Plains crash. When asked in a 1997 radio program if the alien technology he had worked on had come from Roswell, he replied,

**"It was from Roswell, but sometimes I get the idea that some of it may have come from the St. Augustin crash."**

In another videotape interview he stated, **"I was privy to the St. Augustine crash."** (bold by author.)

When asked if this crash could have been located at Socorro, 65 miles away, instead of the Plains of St. Augustin, he said;

**"No, it wasn't Socorro; it was down on St. Augustin."**

He emphasized this point by indicating that a line drawn from north of Roswell through the Trinity site (first A-bomb test site) goes straight through to the Plains of San Augustin. The implication here is that the two crashes (at Roswell and on the Plains) were related. In 1957, Col. Corso was placed in command of an anti-aircraft battalion at Red Canyon missile range (at the north end of 4,000,000-acre White Sands Missile Range area) near the Trinity site. The entrance to the old Red Canyon base is a few miles west of the town of Carrizozo on U.S. Highway 380. I had been working for three years on this project when Col. Corso made these San Augustin disclosures. I could not ask for anyone in a higher position to support my research.

Col. Corso's statement on at least four media occasions, that there was a Plains of St. Augustin crash, added a new dimension to another theory or two that had been considered by researchers for years. The first theory was that two UFOs had collided somewhere in the Southwestern sky, with one coming down north of Roswell and the other on the Plains of San Augustin. The second scenario was that one craft had experienced trouble, with part of it coming down and the remains (perhaps an escape pod) landing on the Plains where Barnett found it. Just after the turn of the millennium according to *Nature Journal*, Taiwanese researchers discovered that gigantic lightning jets can stream upward from thunder clouds as much as sixty miles into the atmosphere. These lightning jets have been photographed similar to spidery tree-like formations. Perhaps the propulsion or guidance system of the UFOs ran afoul of this atmospheric phenomena. Lightning and thunder clouds were reported in a band across central New Mexico on the night of July 1 and 2 of 1947. I tend to go with the first theory of two UFOs going down for whatever reason, as three or four bodies were reported at each site. (Six to eight bodies are too many for a small craft.) Barnett was such a credible witness that there is little doubt about his witnessing the crashed saucer with bodies on the Plains of San Augustin, in July of 1947. We really do not have to tie Barney's find to the Roswell scenario to make a case for a Plains of San Augustin crash, which he shared with some friends and relatives and with Col. Leed.

As to the Plains crash theory, we are not dealing here with the word of your average aging colonel. In Corso, we have a highly-reputable senior officer at the Pentagon who had been assigned to work with actual crashed UFO technical material and distribute it to private industry. After his stint at the Red Canyon base, Col. Corso was given the command of a Nike AA missile battalion in Germany, as well as other commands.

In this book, we have presented several high-ranking military Army and Air Force officers. We also have as a very good witness in retired Col. William Leed, mentioned earlier, who had command responsibilities in the Signal Corps in Korea.[2] Both high-ranking officers, Corso and Leed, were trusted by top generals to lead combat-ready operations in case of hostilities, and were placed in extremely sensitive positions overseas during the Cold War. I doubt that either officer was the least bit delusional or prone to cherish fiction in their personal lives before, during, or after military service.

I am occasionally asked about Ray Santilli's 1995 alien autopsy in connection with the HDPE artifact. Because individual body organs were not photographed separately and cleaned up from whatever body fluid was viewed, I see no similarity between it and the organs extracted on screen during the autopsy. I was surprised at Col. Corso's response to the film, though. He spoke of a group of surgeons who viewed the film and thought that the surgical procedures for an autopsy were correct. But, since autopsy procedures including weighing and measuring were not followed, this group felt that the procedure was more of a dissection than an autopsy.

Col. Corso noted that some aspects in the film did match what he had read and taken notes on: a top-secret alien autopsy at the Walter Reed Army Hospital in Washington, D.C. This lab was financed by Army Research and Development funded through Pentagon sources. Notwithstanding some of Col. Corso's comments, the film is generally classified as a fake among serious UFO investigators.

Col. Corso told his son Philip Jr. just before he died that "the alien muscles were striated or grooved, a condition much more suited to a weightless environment in space." Referring to the autopsy film, he told his son that he was "often asked if the alien autopsy film was real?" He said, "Let me tell you this. I can't testify whether it was real or not, but from an Intelligence Officer's standpoint there are several items that whoever faked this film, if they did fake it, could not know." He went on to tell his son three other things about the alien he noted in his research. "Number one, there was no lymphatic system; number two, the brain had four lobes electromagnetically interconnected [human has two lobes]. One of the lobes was laced with ICs (integrated circuits) [which of course were not in the film]... The other part is that the alien had three eyelids. It had the outside one that was removable, it had a membrane, and then the eyeball."

The eye really intrigued him, and he was very interested in the dark lens removed from the right eye of the alien in the film. The Walter Reed secret autopsy report mentioned this artificial lens found on alien eyes. From the alien artifacts from his file at the Pentagon, Col. Corso said he had taken just such a "light-collecting lens" to private industry for development. He also said, "This later became the basis for night-vision goggles developed for the military." His comment, upon seeing the lens removed in the film, was, "How could they have known?"

Among some UFO investigators, there is conjecture about Col. Corso's validity in his UFO recollections. With some information that the author has researched, Col. Corso's integrity and memory seems beyond reproach. However, this is not to necessarily find fault with other people's compressed and generalized opinions of him. Col. Corso came into the middle of UFO research and immediately ran afoul of some researchers with preconceived ideas and who saw

### THE GRAY ALIEN CLONE

Some of Col. Corso and other's writings have been compiled concerning the gray alien physiological make-up.

| | |
|---|---|
| **Height** | 4 to 4 1/2 feet |
| **Head** | Large by human standards, bulb-shaped |
| **Eyes** | Oddly spaced (Barnett 1947), slanted, large, and black since mid-sixties |
| **Mouth** | Small slit, without lips, no teeth, appears nonfunctional |
| **Reproductive organs** | None visible |
| **Torso** | Small and thin, no skeletal or muscular structure (hairless) |
| **Arms** | Thin, reach to knees |
| **Fingers** | Three to four inches long, some slightly webbed |
| **Legs** | Short, thin, feet usually covered. We are unable to determine whether toes are four or five digits. |
| **Brain** | Larger than ours, three lobes |
| **Lungs/Heart** | Work together as two-piece organ* |
| **Stomach** | None. No digestive or excretory organs. It is believed that if this process occurred, it may be through the skin via osmosis.[3] |

*The artifact (Figure 7.1) may be a part of an artificial heart/lung that Corso referred to in the secret Walter Reed Hospital reports.

some things differently. When all is said and done, all the viewpoints are usually blended by time and become history for subsequent generations to draw upon. Whether we accept the following recollections or not, we are forced to consider it as a recollection of a credible eyewitness until proven otherwise.

Col. Corso said to the various media, that he was a Duty Officer at Fort Riley, Kansas, in mid July of 1947. He related that a caravan of trucks came through from Fort Bliss, Texas, carrying about 30 crates including some long crates with alien bodies from New Mexico. Corso said that while he was in the warehouse with the crates that he looked at the routing manifest, and the bodies were being shipped from Fort Bliss, Texas to the Air Material Command at Wright Field and from there to Walter Reed Army Hospital, Pathology Section. Most UFO investigators assume that the alien bodies from Roswell were flown from New Mexico to Wright Patterson Army Air Force Base near Dayton, Ohio. This of course may have happened to this set of bodies (we believe about four), but the bodies apparently seen by Barney Barnett on the Plains of San Augustin may have provided a second set. We have reason to believe that one or more alien bodies may have been mutilated in the crash. It would make sense to ship some by air and some by ground transport, considering their possible destruction in the event of an airplane crash.

Military records from Col. Corso's service file indicate that Major Corso, Philip J., service no. 01-047-930, was assigned to the General Ground School at Fort Riley, Kansas from Mar. 23 to Sept. 22 1947. He was also at Fort Riley for various duty assignments from 1948 through 1950.[4] This was apparently the beginning of Col. Corso's experience with aliens. Little did he know then, that 14 years later he would be assigned to work at the Pentagon with alien technology and have intimate knowledge of autopsied aliens, at least several of which were being shipped through Fort Riley in 1947. We will consider the possibility in an upcoming chapter that the alien crashed craft may have been taken out by railroad from the nearby town of Magdalena.

**NATURAL MATERIAL**
Throughout this research, I was continually surprised by the abundance of natural materials that showed up in the laboratory analyses. Of course, the foil, the wire, the honeycomb, and large metal pieces were part of a metallurgical technology, and the HDPE artifact was, no doubt, made in a petrochemical environment. Researchers in this field and interested lay people seem to have some expectations that something exotic and dramatic should turn up at UFO crash sites. This may have been the case if the military had not arrived first.

I was impressed not only with exotic finds like the I-beam, honeycomb, and the artifact, but the aliens' innovative use of natural materials we also share on earth. What intrigued me was a series of small surprises and different uses of common material, some in unusual configurations. The natural carnauba-like wax could easily have been from disintegrated candles, but nearby was the T-bar made of the same material, obviously a structural member of some type. Then there was the cellulose material made from plant fiber. This included the wax mentioned earlier, the starch in the HDPE artifact, and the six cellulose layers of the small shoe sole, as well as some plant pitch or lecithin oil in the rubber, which was also a cellulose product.

Two of three kinds of oil found in the shoe sole and on the outside of the HDPE artifact were natural. Lecithin oil then may be a component of the shoe sole. According to *Webster's Dictionary*, lecithin is one of the "waxy hygroscopic phosphatides that are widely distributed in animals and plants." It is also a health food supplement used to promote circulation. The shoe sole was also found to contain fine kaolin-type clay, which is used in ceramics and refractories as an absorbent and filler. Calcium carbonate was found in the small urethane shoe liner. In nature, calcium carbonate is a calcite and aragonite, which comes from plant ashes, shells, etc. It is used in making lime and Portland Cement.

We found 20 or so different items at the site, consisting of dozens of elements and numerous different materials. The metal foil shards, each containing up to 56 elements, were a mystery. We believe up to three-fourths of these elements were trace elements that may have already been in the metallurgy of the metal artifacts. With the exception of the metallic items, the plastic HDPE container and the urethane foam layer in the small sole, everything that could have been made of natural material apparently was. More research needs to be done in this area, and I am unsure whether the use of these natural materials was a coincidence, a manufacturing trend, or an advanced technology with a deliberate blend of environmentally sound materials.

A word seems in order concerning lightness. Everything of interest found in the arroyo seems unusually lightweight. In light of claims of UFOs weighing tons and leaving heavy imprints in the soil, our material was featherweight. The wax pieces, including the T-bar, were very light, as was the shoe sole, and the wafer. The most striking lightweight materials were a number of pieces of foil, and the honeycomb was extremely light for its size. If all or some of these materials are associated with a UFO crash, then weight must have originally been an important consideration. Fewer than a dozen men may have been used to load the craft on a flatbed truck.

**THE SHOE SOLE**
The laboratory was able to determine several interesting things about the shoe sole: the rubber was natural and indications were that it was made of plant pitch with some lecithin oil, pigmented with titanium white dioxide. The upper part of the sole consists of various kinds of woven cotton fibers, which were also embedded in the bottom sole. There are indications that this bottom layer has been vulcanized in at least one step of its construction. One New Hampshire shoe manufacturer wrote, "It seems to have many more layers than is common."

The small size was a different matter. Since the small sole is not over five inches long, the research dealt mainly with children's shoes and their construction. Questions arose regarding the wear pattern on the bottom of the small shoe. Research indicated that we were dealing with a shoe length to fit a child one to two years old. Children of that age change a half to three-fourths of a size every three to four months as they grow. Shoe industry technical literature indicates that shoes of this size never wear out before the next size change. Assuming a child was walking in this shoe, how could a toddler in three or four months wear out the sole in the middle, with no appreciable wear on either end? Another question arises: Why would shoes of nine layers be made for children who would soon outgrow them? Representatives of the Western Shoe Association and Footwear Retailers' Association could not find any data that the small shoe size had ever been made in the U.S.A. or overseas.

After shoe size research, I was left with the conclusion that I was dealing with a mature foot of some type. Nor did the insole pattern fit any known size on the standard foot measuring devices including the familiar Bannock used in shoe stores today. A man from the Nike shoe company near Portland said, "Even if we could find a foot this size to wear this small sole, it is far more sophisticated than we could manufacture for a child who would outgrow it in several months. As mentioned before, we deliberately did not tell people what we thought they might have been looking at, especially things that could be extraterrestrial. We did not have the heart to tell this nice man that the shoe might be alien.

## SUMMARY

In this chapter we considered a variety of topics and ideas. First, we discussed the impact of the craft in the arroyo. There is strong evidence the crashing UFO hit the ground at least twice and skidded 300 feet or so to a stop. We believe the debris pattern suggests an explosion about 150 feet into the final skid. The explosion and "hot environment," probably with a fireball, started to melt previously dismembered body parts, and they were blown out of the craft, as evidenced by the partially melted surface of the HDPE artifact.

Some of the foil shards we found were also charred. The gap we found had no buck brush growth and was obviously the path of the skidding craft. We initially checked the gap for radiation and found none. We eventually assumed that the root structures of the buck brush probably had been removed by leveling equipment, filling in the low trench made by the craft.

The lack of crystalline structure in the HDPE artifact was probably deliberate and not accidental. To function, the artifact needed to be soft, and there is evidence that its softness was probably in its chemical makeup. We use HDPE and its derivatives in many types of minor body applications. It seems this material is impervious to chemicals. On a sadder note, some blown apart body parts were no doubt gathered by the cleanup crew and buried a few yards east of the gap, which will be discussed later.

It was refreshing to see the use of some natural materials, including the carnauba-like wax, which comes from the carnauba palm in our world's tropical climates. Lecithin oil is natural and used in our health industries. Calcium carbonate was found in some urethane in the tiny shoe sole. It is common on earth, as well as natural rubber in the sole. The big surprise was the reliance on cellulose fibers in the shoe sole, which was combined with unidentifiable coatings.

Lastly, and probably the most important person to come into the UFO field in a long time (1993–1998) was Colonel Corso. There was some initial grumbling and credential challenges from top UFO researchers. They had to shift over a little to gradually make room for him, once he was able to prove who he was and where he fit into various government levels. His information was generally accepted, albeit with reservations by some UFO researchers. Like it or not, someone comes along occasionally and pulls us with visible heel marks into a different but better place. Colonel Corso did that for ufology. He will be and has been missed.

## CHAPTER 14
### REFERENCES

1. Leir, Dr. Roger K., *The Aliens and the Scalpel* vol. VI, The New Millennium Library, Granite Publishing, LLC, Columbus, NC, 1998, 1999.

2. Campbell, Art, Private correspondence & interview with retired Army Colonel William Leed.

3. http://www.crowdedskies.com/grey_alien.htm#.

4. Corso, Philip J. 01-047-930 Department of the Army DA Form 66 Officer Qualification Record.

UFO CRASH AT SAN AUGUSTIN

# CHAPTER 15

## PLANE CRASHES, CHICKENS & MISSILES

Some forty miles south of where the Roswell flying saucer allegedly crashed, is the sleepy town of Carrizozo, New Mexico (population 1,459 in 1947). It was here on July 18, 1947, that a terrible airplane accident took place.[1] The reason this event is included in this research is that some UFO researchers think this incident might be related to the Roswell crash. Three other plane crashes occurred 80 miles to the west in the San Augustin area.

First, the Carrizozo story. On July 15, 1947, (one week after the Roswell crash story broke), a P-80 Shooting Star made an emergency landing on U.S. Highway 380 just north of town. The P-80 was one of the first operational jets in the United States Army Air Force. One crash occurred on a routine training flight between March Field, California, to Biggs Field, Texas, near El Paso. It landed because it was low on fuel. Although, additional fuel for the plane had been brought in the day before the attempted take-off, it was discovered that the batteries were low and that a re-charger needed to be flown in. The takeoff was set for the next day, Friday the 18th.

Lincoln County Sheriff Nick Vega and several deputies closed off a mile and a half of U.S. 380 prior to the take-off, which was near the intersection of U.S. 54. The area was sparsely populated, although there was a gas station/small store at the intersection known as Monte Vista. As the plane started its take-off run, the unevenness of the road and the rough surface, combined with some potholes, prevented sufficient speed for take-off. The plane's wheel apparently hit a culvert, causing the plane to veer some 50 feet off the road. It sheared off a tree before plowing into the back of the station's living quarters. The plane then hit the wash and grease house, and as it did, 700 gallons of fuel exploded into flames.

Captain Floyd Soule, the pilot, was immediately killed, and a civilian ex-World War II B-17 pilot, Joe Drake, who tried to rescue him, died the next day from severe burns. Four others were

*Figure 15.1. **Gilbert Amijo** moved to Horse Springs in the early 1940s when his parents bought a store there. He remembers the 1947 light plane crash there, while he was at the Horse Springs School. He recalled all of the children going barefoot and growing what he termed "Alligator feet." He is very proud of his heritage and lived in Reserve, NM, forty miles to the south. He enjoyed showing the author his grandfather's 1872 saddle. Gilbert is one of the many local area people who contributed to this research.*

hurt, including an eight-year-old boy, a lady hit by a pickup truck in the confusion, and Sheriff Vega himself who suffered slight injuries attempting to administer first aid to burn victims. Thirteen people were noted as being at the scene, including several children and two ladies who emerged from the station's restroom shaken, but uninjured. It is believed that many more watched the crash from safer vantage points. A team of Army medical personnel and doctors was called in from Alamogordo Army Air Field bringing blood, plasma and other supplies to the injured civilians. Two of the injured were flown to Alamogordo for treatment, and one to Albuquerque.

On Sunday, July 20, 1947, (two days after the crash), the members of Trinity Church in Carrizozo held a camp meeting for ranch families in the area. Thirteen hundred attended the four-day affair. Prayers were offered on behalf of those who were severely burned in the accident. Rev. Howard S. Pitts offered the following prayer: "As we reflect upon the likelihood that Captain Floyd Soule paid the supreme sacrifice in order that more lives might be spared, we bow in silent tribute to those who dare find the life of the spirit through losing the life of the body. The spirit of self-sacrifice knows no death. May God's abiding love keep the families of Joe Drake and Floyd Soule who have gone to meet their pilot face to face."[2] Area ranchers and townsfolk were quite moved by this tragedy coming on the heels of recently won WWII, while sentiments about God and country were still very strong.

Some UFO researchers take liberties with bits and pieces of information and often jump to conclusions. In this case, because of the highway closure, it was assumed by some, that the government had fabricated the story of a secret plane crash to justify closing the highway so that the Roswell wreckage could be removed in secret to the south towards Alamogordo Army Air Field and/or the White Sands Proving Grounds. It is believed that the highway was closed almost two hours. As far as the P-80 Shooting Star being secret, many of its systems no doubt were secret, but the plane was placed in United States Army Air Force service early in 1945 and had been at many air shows. The P-80 was known as a gas guzzler. In light of this, one thing remains curious. A direct flight from March AFB near Riverside, California, to Biggs Field at San Antonio, does not go over Carrizozo, New Mexico. Did the pilot decide to detour over the Roswell crash site about 40 miles north of Carrizozo? If so, it cost him his life and that of another person.

If one buys into the supposition that U.S. 380 was closed due to the removal of Roswell wreckage, there are some logistical problems. U.S. Highway 54 runs north and south, crossing the east-west U.S. 380 at the intersection called Monte Vista where the crash occurred. North of Carrizozo, State Highway 54 runs much closer to the Roswell crash site through Vaughn, Duran, and Corona, and south to Alamogordo. A few miles above where the Roswell crash is said to have occurred, is N.M. 247, which connects to U.S. 54 at Corona. Also running parallel to U.S. 54 is a railroad. It was originally built by the El Paso and Northeastern in 1899 and was in operation in 1947. The tracks today are owned by the Union Pacific. This is not to say that the UFO wreckage was not transported; it very well may have been. Any transportation on area roads in those years could easily be done at night. Secondary roads in New Mexico, then and now, are very lightly traveled after dusk.

Various ranchers recall two other aircraft mishaps, more in the nature of forced landings. One occurred a few miles east of Datil and a little south, on the Ake ranch. The other was on the Beaver Head Road on the south side of the Plains.

The Ake ranch landing began one morning while Trudy Ake was finishing her morning coffee, and she heard an airplane flying very low. Apparently, an Army C-47 transport was en route from the military base at Fort Huachuca in southern Arizona, to the northeast, with a load of valuable scientific equipment. When the plane experienced engine trouble, it was necessary to make a forced landing on the main road to the Akes' ranch. Pilot and crew were okay, but the plane was somewhat banged up and remained in place a week or so while the ranch crew got to know the Army men who were installing a new engine. For three weeks, people came for miles around to see the plane and meet the ground crew. Trudy said it was branding time, and one of the big military trucks helped pull out a pickup stuck in the mud. She also said that some of the locals kidded the crew about local wildlife, including the coyotes.

In about 1965, another plane made a forced landing. This occurred on a brisk fall day near where a rancher was cutting firewood. He heard the sound of a small low-flying jet and heard a rather noisy landing, then saw a huge cloud of dust. This emergency landing was on the Beaver Head Road some miles south of U.S. Highway 60. It was also due south of the Plains and today's VLA radio-telescopes, which were completed in 1982. The unimproved gravel and mostly dirt road is only passable in good weather and, due to light traffic, gets very little maintenance. The pilot had experienced engine trouble and landed on a long, straight, isolated stretch of the road. Within a few hours, a truck and some air frame mechanics arrived. The pilots were taken back to their base.

**THE HORSE SPRINGS CRASHES**

Two crashes occurred at Horse Springs, New Mexico. The first was a WWII military Trainer aircraft, an AT-6C. The date was February 1, 1944. The second crash occurred in July 2013 when an organization named the Aerial Phenomena Investigations out of Annapolis, Maryland, went down after a badly botched investigation.

Stanton Friedman wrote a scathing two-page rebuke of the API investigation. He said, "It seemed like a hatchet job from the start." Stanton's comments on Antonio Paris's use of the word "likely" is a splendid example of bias.[3] Friedman's humor and logic, combined with an appropriate measure of sarcasm, makes his column, "Perceptions – Plains of San Augustin Crash," well worth reading.

Thanks to the API research, a little more is known about the plane crash at New Horse Springs; at least they gave us a definite date and the pilot's name. Here is what we know. Everyone in the area still alive while this book was being prepared remembered it. School children especially remember because school was let out early so they could go see it. The crash site was about 35 yards north of the New Horse Springs store. The plane flipped over upside down, and the pilot Lt. Eddie Magenheimer was killed. When would-be rescuers reached the plane, the lieutenant was hanging upside down, still in his harness. Several from the store came out, and passersby from the highway also stopped. Someone said schoolchildren were coming up the highway, so several men quickly took Magenheimer down. He was put in the back room of the store and covered with a tarp. The next day, a flatbed truck and some air frame mechanics arrived and an ambulance (for Magenheimer's body). The plane was intact but badly damaged. It was disassembled then loaded for transport.

The aircraft was a North American AT-6C Trainer out of Marana Army Air Field about 300 miles away in southern Arizona. These planes were seen frequently during the war flying over the Plains. Several planes in flight usually worked on instrument flying and navigation problems. The official cause of the incident was listed as "crash-landed due to fuel exhaustion." The AT-6C had a range of 750 miles. When the tanks were full, it held 111 gallons of aviation gas. Perhaps the tanks were not completely full or an unknown fuel leak occurred. Lt. Magenheimer died a hero to the Horse Springs residents. As he was coming in to land on Highway 12, which was gravel in those days, a pickup truck with a rancher and two children pulled out from under the New Horse Springs store and post office overhang, going the same way. The plane's landing speed was far greater than the slow-moving pickup. The lieutenant made a quick decision that probably saved the lives of those in the pickup, but it cost him his own. He turned into the side pasture north of the store, and his wheels hit a chicken house roof. His plane completely flipped over and came to rest upside down. Lt. Magenheimer was 22 years old. He was from Mooresville, Indiana.

In 2013 two men from the Annapolis-based API organization came to the Plains crash site. They represented themselves as a MUFON Star Investigation Team. We gave them every courtesy and cooperated with them in any way we could. After about three weeks, we received their report titled, "How Physical Evidence, a Detailed Investigative Process, and Good Detective Work Solved the 1947 San Agustin Mystery." The API team determined we had found the wreckage of Lt. Magenheimer's AT-6C Trainer.

Besides the lack of independent testing of metallic metal shards that they found, plus the ones we sent, which they agreed to test and didn't, there were very serious problems with their conclusions. The crash debris we found over a nine-year span in a 400-yard-long swath was two valleys away over a high ridge and two-and-a-half miles from the store. The official government report reads, "crash landed due to fuel exhaustion." The New Mexico Highway 12 where the pilot attempted to land is very straight in front of the store for over a mile. This in no way resembles the rocky terrain, juniper trees, and soft sand of the

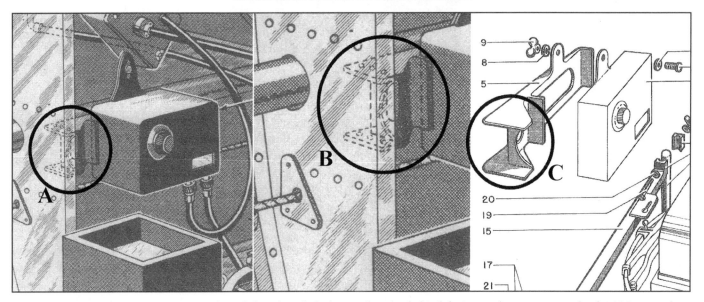

*Figure 15.2.* **Misdiagnosed I-beam.** *A, above left and circled, shows a bracket behind the intervalometer as seen by the API researchers in the "Pilot's Flight Operating Instructions" manual on page 35. B, middle and circled, is an enlargement of the bracket end shown in A. In C, right and circled, is the bracket end shown in a North American Aviation parts catalogue from 1942. The part shown was labeled as a bracket for an aerial camera timing apparatus, Part No. 54-73116.*

arroyo where the API was convinced the wreckage, including the I-beam, came from.

The entire premise of the API investigation was that an I-beam found in 2012 was from the crashed Trainer. Our ICP-MS tests and isotope analysis indicated (see Chapter 6) that the metal was not of earthly origin. By 1939, all aircraft metal for military planes manufactured in the U.S. was standardized. An alloy known as 24ST was extruded for military aircraft structural purposes. [4] No I-beams of any size were used in the AT-6C Trainer and earlier models dating back to 1935. Also, all military aircraft structural components were coated with a zinc oxide paint to prevent corrosion. If AT-6C parts were scattered near the UFO crash site, some of this greenish-yellow paint would have been obvious. [5]

How then could an I-beam be found in an AT-6C that fooled the API team, and as they claimed, also fooled two aircraft archaeological experts they claim to have consulted. They misread a part in a 1945 *Pilot's Operating Instructions* manual as an I-beam.[6] When enlarged, it was clearly not an I-beam. (See Figure 15.2, B). We found the actual part in a 1942 AT-6C and similar models parts catalogue. [7] This is where their investigation began to fall apart. It was obvious to us that they were more interested in challenging our I-beam found a year earlier (of possible extraterrestrial origin) than objectivity researching the plane crash in 1944 or our materials. Their "I-beam" was a simple $4\frac{1}{2}$-inch-long bracket with holes for aircraft lightness, was behind a timing device for an aerial camera. Any amateur draftsman could clearly see the difference.

There was more. The API team disregarded early witness accounts, including Barnett's, as being "hearsay" because the information was thirty years old. Did they accept the word of Jesse Marcel, with no physical evidence from the same time frame? They disregarded some Air Force intelligence information as "local lore." They ignored retired Colonel William Leed's corroborating account (from Pentagon sources) of Barnett's story. We were chastised for not following "chain of evidence procedures." We used appropriate and prudent record procedures for our situation (see Figure 5.4). Behind the honeycomb, one can see a plastic bag with writing on it. We designated a "bag and tagger" and photographer for our finds. Writing can be plainly seen on the plastic bag that corresponded to an entry in a notebook page for the artifact. But they did not ask for any documentation, and none was offered. Some of our documentation contains confidential material. We were not doing any litigation or crime scene investigation and did not worry about fingerprints. We knew the ICP-MS process was thorough, and we also knew our material would be pulverized, placed in a liquid, and atomized, then sprayed into a 6,000°F chamber. Fingerprint contamination with this process seemed remote. As long as we are considering "chain of evidence procedures," does this include distorting a photo in their article to make the photo evidence fit API's theory that the marking on the I-beam was the same as what appears on aircraft fastener hardware? In an early report sent to Stanton Friedman and posted online, the triangle image appears as an equilateral triangle instead of the actual aspect ratio as shown in Figure 6.2 in this book.

Another criticism of our procedures was that we did not vet our material to the scientific community. Unbeknown to the API team, we wrote a letter to the then national director of MUFON (Mutual UFO Network), Mr. David MacDonald in August 2012, ten months earlier. Accompanying the letter was a packet of materials dealing with the scientific research on this project, along with some photos. Mr. MacDonald turned the material over to his Science Review Board. We thought it was appropriate he started with the MUFON science team. But we did not know whether API and its MUFON-trained investigative team were connected with the Science Review Board or MUFON leadership, so we decided to cooperate with them. In their article, the API completely disqualified our scientific advisor Steve Colbern as not objective because he believed in UFOs and was associated with the UFO community. Exactly the same associations as their own. Mr. Colbern graduated from UCLA with a degree in Chemistry and is a well known materials analyst. Some of the API directors' scientific credentials came from a correspondence school.

When details of this "investigation" were sent to MUFON leadership, the following statement was received from Maryland's State MUFON Director. He said, "I trust you are aware that this investigation was handled solely by Antonio [Paris] and his Aerial Phenomena Research Investigation team (API) and that MUFON was in no way involved."[8] We were certainly relieved that MUFON was in no way involved and we believed they were also. Mr. Paris had some 15 aluminum samples from the site. After Mr. Paris's negative report he was asked to return the samples and we received an October 2013 email, "The material was analyzed and it is concluded as being 100% aluminum from local trash including aluminum from beer cans. We therefore recently destroyed and discarded the material."[9] This is the same honeycomb material featured in Chapter 5, some of which we sent to him. Steve Colburn, our chemist/ materials analyst, concluded that this metal was largely composed of aluminum, chromium, iron, and magnesium. (The reader can see in a shadowbox on pg. 37 the metallurgical content of W-1102.) Steve did not think the technolgy to blend these major metals and 30 or so minor elements was available in 1947. Is it possible that the API investigation concerned an alien beer can or were the strange isotope combinations a later Earthly development? After this botched report and ludicrous conclusions, Mr. Paris was no longer associated with MUFON's investigative units.

There was no recollection of any UFO crash by Plains residents in those days. But all I talked to did recall the plane crash at the New Horse Springs Store. Josie Argon recalled it but as a first grader, she remembered little. Gilbert Amijo recalls the crash because of the egg shortage. The postmistress kept a large flock of chickens where some locals bought their eggs. Gilbert said the plane took off the roof of the chicken house and there were chickens all over the place, "some injured, squawking and a' flapping" in the pasture. The postmistress gave them to anybody who could catch them. Gilbert and a friend got two. He said they had plenty of chicken in the school lunches; everybody in Horse Springs had lots of chicken the next several days. They had fried chicken, boiled chicken, chicken and dumplings, creamed chicken, roast chicken, and lots of feathers, but when that plane took the chicken coop roof off, they had no eggs for a long, long time.

## MISSILES

There were three publicized types of missiles under development at the White Sands Proving Grounds in 1947–1948, 150 miles southeast of the Plains; air-to-air (AAM), surface-to-air (SAM), and surface-to-surface (SSM) missiles. One of the air-to-air missiles under development in 1947 was the GAR-1, which was a missile later designated in the Falcon family. One of the surface-to-air missiles under development in those days was the Boeing GAPA (Ground-to-Air Pilotless Aircraft).

In an article in the *Albuquerque Journal* of August 24, 1947, the GAPA super-speed rocket was said to be a "new secret missile that chases planes." The most well-known surface-to-surface missile in 1947 was the V-2, which was acquired from Germany after World War II. A launch was described in the October 16, 1947 issue of the *Adair County Times,* where it was stated that the missile reached an altitude of 100 miles at a speed of 3,600 miles per hour. The Germans had launched the V-2 at targets over 200 miles away. The V-2 experiments paved the way for America's first intermediate ballistic missile, the Jupiter.

Another German-developed missile being tested at the White Sands Proving Grounds was the famous "buzz bomb." It was designated by its German developers as the V-1, with the United States Air Force designation of the JB-2. This weapon had a range of some 150 miles, cruised at a maximum of 3,000 feet with a speed of 559 miles per hour. The JB-2 had enough range to reach the southeastern edge of the Plains, but its 3,000-foot ceiling would probably not have been capable of clearing high peaks in the Black Mountain Range, Diamond and Paloma Peaks.

Some research on the use of aluminum honeycomb structures indicated that the JB-2 had some experimental honeycomb configurations in its short, stubby wings. This was the only use of an aluminum honeycomb in aircraft or missiles we could find prior to 1950.

*Fig. 15.3.* **Military ammo casings and a military presence at crash site.** *Larger 50-cal. ammo casings found nearby, left over from aerial gunnery practice during WWII. Smaller 30-cal. casings had been used in an M-1 Garand rifle. It is doubtful such casings were fired at site, but probably dropped out of a military pack. Small can rings meet size specifications of military K-ration cans. What we believe is a military bivouac was also found. Mlitary canteen (flattened by a heavy vehicle) was of WWI (1918) issue - probably reissued - in early years of WWII. Canteen was found in 2005 by an archaeologist walking at site.*

Captain James McAndrew was an intelligence applications officer assigned to the secretary of the Air Force. In 1997 he wrote a book called *The Roswell Report: Case Closed.* Captain McAndrew commented on missiles launched from the White Sands Missile Range in 1947. Speaking of the famous Roswell crash site and the lesser-known Plains of San Augustin site, he said, "These (missile) projects were well-documented, and none of these missiles landed near either of these crash sites."[10] There seem to have been no V-2 or buzz bomb crashes on the Plains in those days, either deliberate or accidental.

There is some conjecture about early UFO crashes being attributed to missiles. There is no doubt that missile experiments were underway by several countries after WWII. We had developed operational radar systems shortly before Pearl Harbor. But detecting aircraft and locking them into effective missiles was more than a decade away. As to effective systems working together, development of guidance, propulsion, and radar systems at the same time led to long delays.[11] The U.S. Nike Ajax missile was not operational until 1952. Colonel Corso (as previously mentioned) was an early missile battalion commander in Germany and we believe the Nike Hercules surface-to-air missile was deployed in 1958 in his unit's arsenal. Early interception by our aircraft often ended in disaster, as in the famous Mantell case of 1948.

We occasionally get a tongue and cheek acceptance of the metal pieces we found at the crash site "as probably from some missile or other." There was a great deal of friction over the government acquiring land for the White Sands Proving Ground during and after World War II. They were extremely careful about missile firings over any areas where there were people or livestock. It was jettisoned boosters that the military was concerned about.

However, later in the 1960s, missiles were being fired across the top half of the Plains about 60 miles to the north of the crash site. In these years, intermediate range missiles were being tested that flew in a high arc diagonally over the north end of the Plains. In this area, U.S. Highway 60 bisects the north end of the Plains for about 25 miles between Datil and Magdalena. The military had a radar site in an old quarry along Highway 60 outside of Magdalena. It, of course, could track the high missiles but was too low and surrounded by hills to be any buffer or threat to UFOs that may have occasionally flown over. Some of these 1960s firings are blended in some foggy memories and are associated with UFOs, but these missile firings were a separate event.

There were some disruptions to local lives in those days, not necessarily so much from the missiles themselves, but jettisoned boosters coming down, which could conceivably land near or on ranch buildings. Ranchers remember 15 or 20 of these firings in the early '60s. The government did everything they could "to get us out of the way," one rancher recalled. "They would give us a week or ten days to make other plans and move our livestock and horses in case there was any trouble." A rancher went on to say that at government expense, those living in the missile track could go to a motel or hotel. They would also pick up meal expenses for a day or so. At first "It was a nice, novel idea and we had a good time seeing all of our friends. Most of the gals would gab in some room or other and most of the men could be found at the nearest bar," he said. A tired joke usually circulated at these informal get-togethers. It went like this: "If the government wants our land, all they would have to do is drop one of them missiles on the bar or motel and get a bunch of us." Apparently when the missile was passing over, they closed highways for a time. Another rancher recalled being "held up" for an hour or so on Highway 60 near Datil. I asked him where the missiles were being fired from. He said, "Somewhere near and south of Green River, Wyoming." Another ranch wife indicated that it was disruptive, but she said "we had a chance to get out and make acquaintances and see our neighbors. The motel and restaurant people loved it." Apparently the firings tapered off by 1968.

One day I ate my lunch in the shade of a large rock on the edge of the Plains some distance from the gap. The rock and a scrubby juniper or two provided the only natural shade I could find. I had brought with me a metal detector borrowed from a rancher. While testing it, I ran it over the ground near the rock and immediately picked up a sharp beep and the gauge's needle jumped. When I kicked away the dirt, there on the surface was a 30-06 brass rifle cartridge. I didn't think much about it because 30-06 is a standard hunting rifle size.

After lunch I tested the area further, only to find some small tin can lids and some rings that resembled those found on an old Spam or sardine can, or possible from Military K-rations. Nearby, a really loud beep sounded, and this resulted in a whole clip of empty, previously-fired 30-cal. brass shell casings (see Figure.15.2.) There were more of these nearby. While in the Navy, I had fired an M-1 Garand rifle on a rifle range. Additional research on my finds indicated that the clip of eight empty cartridges was for that rifle. I later asked a rancher who had served in the Korean War, if any of these rifles were ever used for hunting in the area. He was not aware of the use of this rifle for that purpose, and further recalled the weight of the rifle as about ten pounds. He also recalled that it was semi-automatic and probably too heavy for hunting.

During my many trips to the Plains, I also found some 50-cal. empty brass cartridges, a clip or two, and several slugs. These heavy machine gun cartridges remained a mystery until I later saw a half-bucketful of this material in the garage of a Socorro geologist. He said he had found these in the mountains and on the Plains over a number of years, and that they were from Army Air Force aerial gunnery practice. He also told me that during WWII, bombers from Kirtland Air Force base near Albuquerque occasionally flew very low over the Plains, spooking the cattle and shooting at antelope.

I did considerable research on the 30-06 ammunition casings previously mentioned, and learned that the base plates on all ammunition casings have code numbers and letters. The markings include year and place and manufacture and loading. The large 50-cal. cartridges used in aerial gunnery practice, were made at several locations in 1942 and 1943. I zeroed in on nine of the ten 30-cal. casings found; the eight in the M-1 Garand clip were made in 1942-43 at the Denver Arsenal, and one was made in Pennsylvania at the Frankford Arsenal. This put the 30-cal. cartridges manufactured and loaded within four or five years previous to the 1947 Plains UFO crash. This ammunition would easily have been available to any area military unit, including army personnel that Barney saw at the crash site.

A retired U.S. Army rifle range sergeant told me that military personnel who had used weapons were periodically required to go to the rifle range, but they had to account for each shot. The easiest way to do this was to pick up their brass (expended cartridges). He said at some ranges, they returned them, as soon as they were done firing. At other ranges the men kept them and turned them in later. I could only surmise that my shady lunch spot had perhaps been a bivouac for one or more of the military people who Barney said had arrived at the site. Perhaps the cartridges had fallen out of a bag or something.

As I talked with area ranchers, I asked if there had been any military activity on the Plains during WWII. The indication was that there had not been any, but some older ranchers did recall occasional military convoys on Highway 60 between Socorro and the Arizona line. One or two remembered the low-flying planes over the Plains during the war years, apparently including the training flights from Marana Army Air Field.

Some aircraft debris from World War II has been found on the Plains. One rancher found a discarded wing tank, and another from the North Lake area found a plastic observation bubble from a B-25 bomber (later used for incubating chicken eggs!). On October 16, 1942, a B-26 bomber crashed into the North Baldy Mountain (south of Magdalena), killing a crew of nine. It is said that wreckage can still be found there. There are signs of military activity in various parts of the Plains, but the only known association with any military recovery of any kind was in the case of the rancher (previously mentioned) who, in the 1960s, chased a NASA recovery team off his property for not obtaining prior permission to come onto his land.

**WHERE ARE THEY NOW?**
What of Vern and Jean Maltais, Alice Knight, Harold Baca, Fleck Danley and Andy Key? And what happened to Wesley Hurt, who vigorously contested Dick's Bat Cave research. Danley died in 1997 at the age of 92, still supporting Barney's story after 45 years. In 2012, Harold and Martha Baca, Barney's old neighbors, were still in Socorro running their large gift shop on Highway 85, Socorro's main drag. Jean Maltais died in 1999, and Vern passed away in 2001, he gave nearly 60 years of service to his country. He was in WWII and the Korean War and served over 40 years as the Veterans Service Officer of Beltrami County, Minnesota. Andy Key, the wrangler and ranch hand who worked near the arroyo in 1947, died in the mid-1990s at the age of 91. He had retired with his second wife Mildred, to Gravette, Arkansas, and is buried near there. A quiet man, he said little about the "missile recovery" he had helped the army with on the Plains after WWII. Wesley Hurt recovered from his disappointment of not being able to do his Ph.D work on Bat Cave. He eventually became quite well known, earned his Ph.D. from Michigan's first Ph.D. class in 1952 and was a prolific writer. When contacted in 1997 and asked if he recalled Herbert Dick, he said, "Yes, I remember Herbie, and he was no friend of mine." Hurt passed away in 1998.

In 2012, Alice was still leading an active life in Dalhart, Texas. Her marriage to J.B. sadly lasted only seven short years when he was struck down by a heart attack in 1955. Alice is a gracious and stately lady. She has a ready laugh and a warm personality. During one interview in Dalhart, she looked out a window as if looking back in time and said, 'I've been alone all these years...I never remarried, you know."

What happened to the members of the extremely productive digs in 2004–05 and 2011–12? They have scattered to the four winds. Chuck Wade, the architect of the last two digs is retired and living in Gallup, New Mexico. At this writing, Chuck, a man of many interests, including UFOs, was also exploring aspects of alternative energy and planning for the next UFO film festival in Gallup. Most of the Navajo crew are still in the land of their ancestors in New Mexico. Benjamin Shirley, who found the first large piece of the UFO sadly departed this earth in 2005. He, nor we, did not realize the importance of his find until the metal isotopes were analyzed in a laboratory. Dick Shirley, his brother, was working in 2012 in the Gallup area. Merle Lawrence was known to be living and working in Albuquerque. Murff Tilden completed his journalistic training and has written several published pieces. Cara Fay was living in Arizona and had once worked with Walt Disney studios in post-production. At this writing she has combined video graphics and her interest in the internet with a quest for spiritual awareness. She helped find the all-important honeycomb that fruitful first day on the 2011 dig. None of the extraterrestrial isotopic research could have been completed without the assistance and expertise of our Chemistry and materials analyst Steve Colbern. He at the time of this writing was living in southern California doing materials research and consulting work. He was also the chief scientist listed for A & S Research, Inc.

*Fig. 15.4. Gerald and his father on the outskirts of Albuquerque in Fall 1947. They and several members of their family came upon the UFO crash in early July 1947.*

How did Gerald's father, Glenn Anderson, fare at his new job. Glenn brought his family west in the summer of 1947 from Indianapolis to Albuquerque. We learned earlier in Chapter 5 that he had taken an important job with the Sandia Corporation in 1948. They were prime contractors to the Atomic Energy Commission. From the *Sandia Corporation Lab News* in 1966,[12] we learned that Glenn Anderson was retiring. The article related "he is a model instrument maker, doing precision machining in Project Section 4253-1." Glenn is quoted as saying "I'm going to make a trip to California to see my new baby granddaughter and I won't have to hurry back." He was an expert tool and die maker with Sandia for 18 years. Gerald also worked in the mid-sixties for a private security contractor and was an armed security guard at Sandia's Lassetter Range. Gerald is now retired and living in southwest Missouri. His dad, Glenn, passed away in 1978.

Despite the odds against it happening, it did. Gerald's

complete 65-year story is now known and he is the only living witness alive at this time who saw the immediate aftermath of the Plains crashed spacecraft. Long before he became re-included in this research, we had discovered the "fill" in Gerald's "furrow" that the craft made coming into the arroyo. The "explosion in an aluminum factory" he described turned out to be the foil shards buried in the sand that Open Mind Organization had analyzed by Steve Colbern in 2010. By far the most important recollection by Gerald was the damage to the bottom of the craft that he told Kevin Randle about in the 1990 interview. He said, "the underside had sustained a lot of damage." No other source including Barney Barnett, had mentioned this damage. Pieces of the craft's bottom were strewn along a quarter-mile of the crash path. (See Figure 5.1) Gerald co-led two very productive digs in 2011 and 2012 and recovered important crash debris. From these finds, researchers and scientists were able, with a reasonable amount of certainty, to reconstruct and understand part of the craft's anti-gravity system.

And what happened to the well-known, prominent UFO writers that gave Gerald Anderson such a hard time? If you have read much of the UFO field, you will recognize the men as Kevin D. Randle, the lead writer, Donald R. Schmitt, the aspiring medical illustrator, and their researcher Thomas Carey. To his credit, Schmitt finally admitted he was a Hartford mail carrier and went on to pursue his interest in UFOs with some distinction. A man with less fortitude would have chucked it all and might have disappeared into obscurity. The study of ufology is much richer now than it would be if Schmitt had left. He has made a credible name for himself. Kevin Randle split away from the team in 1995, and he and Don Schmitt did not speak to each other for some time.[12]

Kevin Randle, with a budding military career in his life, had seen duty in the Viet Nam and Iraq wars. He was in active duty and active reserves of the U.S. Air Force. After 9/11, he joined the Iowa National Guard and was deployed to Iraq in 2003, pulling off two tours of duty, some of it in combat. He retired a lieutenant colonel from Air Force intelligence in 2009.

After Randle left the UFO writer's team in 1995, Don Schmitt partnered up with the researcher Tom Carey in 1998 and produced quite a credible book, published in 2007, *Witness to Roswell: Unmasking the 60-Year Cover-up*. It it a straight story about those who knew of and participated in various aspects of the Roswell UFO crash. Many witnesses testify as to their role and involvement at the air base USAAF headquarters and at the debris field clean-up site. If the actual site where the crashed craft ended up was known to any civilian UFO researchers, nothing of any substance has surfaced. Don Schmitt did continue his education and graduated *cum laude* from Concordia University, and in 2013 was living on a small ranch northwest of Milwaukee, near the small town of Holy Hill, Wisconsin. We wish him well. Kevin Randle last we heard was living in Cedar Rapids, Iowa. Two and a half hours drive to the northwest is the town of Humboldt, Iowa. Barney Barnett, who Randle tried so hard to discredit, gave the valedictorian address to his Humboldt College graduating class in 1914. He rose to the rank of Captain in less than two years in his WWI service.

I have met Don Schmitt but have not had the privilege of meeting Kevin Randle. Both are good writers together or separately, but neither have been remotely involved in productive research bringing forward the Plains crash story. In passing, both have been proactive in denying the Plains crash event. The hype, recognition, and status in the earlier days of being a Roswell expert has, we believe, had something to do with this. Their earlier conflict with Stanton Friedman and later Gerald Anderson did little to give the Plains site any credibility.

I will say that their leaning away from the Plains crash while they were working together prior to 1995 and separately after that point did not do any specific damage to the Plains research. But their influence on other writers, researchers, and conference event planners was significant. One can still find in their 1990s to 2005 or so UFO books, magazine articles, and Internet blogs footnoting the speculations of Barney Barnett finding the Roswell crash north of Roswell with some students, Barnett driving to the Roswell area while telling his wife he was going to Pie Town, due west of Socorro. One well-known writer made reference to Barnett's diary not taking the time to find out it was his wife Ruth Barnett who wrote and kept it. This was probably not Randle or Schmitt's fault but it showed how pervasive their influence was on other writers and researchers. We were not aware of nor did we see any research as to the specific geographic area Barney Barnett was assigned to in his Salado Soil and Water Conservation District. Had there been a little more diligent research, these early influential writers could have found out that Barnett's district was 70 miles north to south, and 72 miles east to west in a huge block of land due west of Socorro. This was his assigned area; this is where he worked and where his responsibilities were. The over 5,000-square-mile area included the Plains of San Augustin. The district was established by a referendum vote of the area landowners eleven days after Pearl Harbor, December 18, 1941.[13]

The basic investigation on Barnett's work area, his life, and discovery were not thoroughly researched. Like many writers they made or influenced other researchers to make calculated guesses, which in the end recognized, by inference or design, Roswell as the only valid 1947 New Mexico UFO crash. To add insult to injury, it seemed they did not object to various bogus claims of this person or that trying to insert themselves onto the Roswell bandwagon, which they and assorted Roswell researchers and followers did much to create. Like it or not, whether one accepts it or not, the extraterrestrial material shown with valid scientific research in this book indicates there was an out-of-this-world UFO crash on the Plains of San Augustin. This is physical evidence that has not emerged to any great extent from the Roswell crash research.

Other writers who came after the Randle and Schmitt 1990s books and the 2003 *IU* article that Kevin Randle helped author appeared to have covered the possible Barnett scenarios. They had! That is all but two:

- The Plains crash happened in early July 1947, and

- The Barnett story as told to his good friend Vern Maltais and his boss Fleck Danley was true.

## CHAPTER 15
### REFERENCES

1. *Lincoln County News and Carrizozo Outlook*, July 19, 1947, Era B. Smith, editor and publisher, Carrizozo, New Mexico, pp. 1–2.
2. Ibid.
3. Friendman, Stanton Column, Perceptions – Plains UFO Crash, MUFON Journal #545 September 2013 p. 10-11
4. *Handbook of Instructions for the Structural Repair of the "Texan" Trainer Airplanes, Series AT-6A, AT-6B, AT-6C, SNJ-3, and SN-J4.* North American Aviation, January 18, 1943.
5. Ibid.
6. *North American Aviation, Inc. Pilots Flight Operating Instructions for Army AT-6C British Model Harvard 11A Navy Model SNJ-4*, January 5, 1945, p. 35.
7. *Preliminary Illustrated Parts Catalogue AT-6C-1-NA.* North American Aviation, Inc. Dallas, Inglewood, Kansas City, November 25, 1942. Report No. NA-5578, p. 123.
8. Campbell, Art. Private correspondence with Don Borton, MUFON Director, Maryland, July/August 2013.
9. A personal correspondance from Antonio Paris regarding return of samples, *October 13, 2013*.
10. McAndrew, James, Capt. *The Roswell Report Case Closed*, Barnes and Noble, 1997, p. 16.
11. Bezick, Scott M., et al. *Inertial Navigation for Guided Missile Systems.* http://jhuapl.edu/techdigest/td/td2804/Bezick.pdf
12. *Sandia Corporation Lab News*, vol. 18, no. 15, July 29, 1966.
13. Clary, David A. *Before and After Roswell: The Flying Saucer in America, 1947-1999*, Xlibris Corporation, 2001, p. 166.
14. *The Socorro Chieftain*, Thursday, August 14, 1947, p.4.

## RECOMMENDED READING
MUFON Journal, Stanton Friedman Column "Perceptions" – Plains UFO Crash, September 2013

# CHAPTER 16

## BRINGING IT ALL TOGETHER

### THE LAND TRANSFER

Late in this research I learned that a rancher who was a large landowner ceded some parcels of land in and around the Plains of San Augustin to the United States government. Interestingly enough, this was the same rancher who owned the land around the arroyo site in 1947. There were two transactions: April 30, 1942, and November 7, 1947. I was able to locate these land plots on a 7.5 minute series topographical map from the United States Department of the Interior. The 1947 transfer was very large, consisting of 1,485 acres. Apparently, the rancher traded his land for land owned by the government. None of the land he traded in late 1947 was near the arroyo site I worked in this project. However records indicate that about the time of the crash, Barney Barnett was doing some surveying for the rancher, prior to his transactions with the U.S. Dept. of the Interior (today's BLM).

In 1947, Andy Key was living with his wife, Phoebe Ann on the edge of the Plains He was a hired hand for L.B. Moore, an absentee rancher, who owned and controlled access to the crash site. On June 10, 1948 his wife died, and he remarried about six months later. As the second family was growing up, their dad occasionally mentioned a missile he had helped the Army recover at the ranch site on the Plains sometime after the war.

In April of 1948 Andy and his wife Phoebe had their first child. The baby was named Joy Ann; the tiny infant only lived one day. About two months later, Phoebe became very ill. On June 9, 1948, unconscious and in poor condition, she was rushed from the L.B. Moore Ranch, where they were working to the Socorro Hospital. She had been under treatment for a brain abscess for over a month. A brain abscess is usually caused by a bacterial infection in the subdural space between the skull and the brain. This may have been brought about by a sinus or ear infection. This may have been a result of some undiagnosed complications surrounding child birth, two months earlier. Phoebe Ann Key was born on Christmas day in 1930. She was 17 1/2 years old when she died. We spend a little more time on these events than may seem necessary, but Ann Key's death became a sad benchmark to Andy Key's recollections of "a missile crash" and cleanup on his ranch owner's property.

I interviewed the second Mrs. Key (Mildred), who had married Andy and was now living in northwest Arkansas. Mildred was Phoebe Ann's best friend. Andy spent the Christmas vacation in 1948 at Hayden, Arizona (after his young wife's death), and there he renewed his friendship with Mildred. They married in January of 1949. I also spoke with a son of Andy by his first wife, who lives nearby. Both remember Andy having something to do with a "missile cleanup or recovery" on the property where he worked about a year before his first wife died. He may also have been offered incentives for his cooperation similar to Mack Brazel's situation at Roswell one week later. Mrs. Ann Key's (Andy's first wife's) funeral notice appeared in the *Socorro Chieftain* on June 24, 1948. It is believed that the Plains crash occurred the first week of July 1947, one week short of a year earlier.

*Fig. 16.1. **Andy Key** was a wrangler and ranch foreman who lived close to Barnett's crash site discovery. He revealed to his sons and second wife that he was called upon to help clean up a "missile" crash on the ranch about a year before his first wife Phoebe died. A search of government records indicate there were no missile crashes on the Plains.* ***Phoebe Ann Key, inset***: *A brief obituary notice of her death appeared in the* Socorro Chieftain *on June 24, 1948. This helped date the UFO crash to early July 1947, same time frame as the Roswell crash.*

Did Andy Key help recover the crashed saucer Barney found? Was he involved in one or two of the possible cleanups? Was he asked to keep the incident quiet, as was Barney Barnett? I think it is entirely possible for Andy Key to have been involved in the recovery and/or cleanup. He and others probably were called in to help clean up the site after the saucer had been removed. His work would have included operating equipment to level the ground and picking up loose material. He worked for L.B. Moore, the rancher previously mentioned who controlled the land around the arroyo and had a home on one of the back roads to the site. Moore was an absentee landowner. Andy Key and his new wife lived in Moore's old house and watched over a few head of cattle and took care of the property. The loading pit, which we found, was less than a quarter-mile from his house. Andy may not have been in on the initial saucer recovery, but living near the site and in the employ of the rancher who controlled access to the land, he would have been a logical one to assist afterwards. According to Andy's second wife whom we met near Gravette, Arkansas, he said he had been asked to help clean up the site in the arroyo. He may have been complicit in keeping the military presence at the site quiet. A mile-long, isolated driveway would let the military come and go. The 'dozer-cut road and loading pit were connected to this driveway. The 2001 and 2012 dig crews found the remains of old canvas army cot frames among some juniper trees near the crash site that might have been associated with a military bivouac.

In the summer and fall of 1947 to the southeast of the Plains, just over the Black Range and San Mateo Mountains, a lot of controversy was brewing. The government was attempting to take more land for the White Sands Proving Grounds by doubling it from 1,250,000 acres to 3,263,000 acres. A huge amount of land had been taken in 1942 also, and affected the grazing lands of over 100 ranches. There was talk of the government condemning this additional land with inadequate compensation to the ranchers. The government eventually took 4,000,000 acres.

In the middle of this dispute was the United States Department of the Interior, through whom Andy's employer was negotiating a land trade. With all the controversy over land status in New Mexico, it would have been prudent for the rancher to say as little as possible about any UFO crash on or near his land, lest the status of some of the land be changed. For all we know, his silence may have helped move the land trade along.

**SECURITY**
Keeping secret a crash site on the Plains would have created different problems because the site was no doubt visible from a road. This presented special security considerations from the onset. First, the site would be cleared of curiosity seekers, the archaeologists, Barney, and any others who may have been attracted there by the commotion. Anyone today seeing a shiny aluminum-like surface where it did not belong would think plane crash! Are there survivors? And can I help? The same thoughts and concerns would have gone through the minds of those who first saw the sun glinting off an object on the Plains in 1947.

Assuming the military arrived soon after the civilians were there and the civilians were herded away, the next step would be to cover the reflecting surface or block it off from view. This could have been accomplished in two ways: (1) Moving military vehicles close to the downed UFO, (2) Covering all or part of the UFO with a tarp or other suitable material. Perhaps a camouflage net had been brought along or a large tent to drape over the craft. Even standard military ponchos weighted with rocks could have been used to help break up the glinting surface visible from some distance.

Once the site was secured and the curious herded away, the next problem would be getting the flying saucer out of there. Since we do not know the weight and actual size, it would be hard to judge what equipment was needed. My best guess is that the equipment was dispatched to the area and the whole operation was done at night. I also believe that one or more vehicles were used and the UFO was loaded onto a truck suitable to highway travel. A rancher showed me a loading pit about a mile from the arroyo site where I was working. A pit such as this is a simple cut by a bulldozer into a bank with a flat surface or road on top. It was obviously hastily dug and has seen little use. Cattle ranches have little need for heavy equipment and seldom need to load or unload it. Livestock, when shipped, were traditionally herded into loading pens with wooden ramps going up into the high truck beds. These can be seen every few miles along secondary roads throughout the West.

**TRANSPORTING THE CRAFT**
One of the first priorities at the crash site in addition to beginning a clean-up would be getting the craft out of the arroyo. A year later in 1948, a very large UFO crashed near Aztec, New Mexico and was dismantled before being taken out. The craft that crashed in the arroyo was considerably smaller than the Aztec vehicle and may have been the first post-war UFO event. Because the military may have been inexperienced at dismantling UFOs in 1947, we believe the Plains craft was hauled out in one piece. We think it is possible a large cradle was constructed on one side of a low flatbed truck and the craft was stood on edge, loaded, and leaned against the high, sturdy structure. Even at an angle of 50 or 60 degrees, a 28-foot craft would have been over 20 feet tall on the truck. As mentioned earlier, there was some evidence that a bulldozer had been unloaded and it may have cut a crude road for the truck as it left with the crashed craft.

New Mexico Highway 12, which ran along the west side of the Plains, was straight (although unpaved) until it reached U.S. Highway 60 about 25 miles to the north. Here the road was paved and also straight for 30 or so miles across the top of the Plains until it reached Magdalena. If the huge load was traveling east towards the nearest military bases, it would have to initially go the 26 miles between Magdalena and Socorro. U.S. Highway 60 was relatively straight east of Magdalena for a few miles, then began a descent of over 2,000 feet in the next 20 miles. The road had many switchbacks and sharp turns in those days, and a load of this magnitude would be extremely difficult and unsafe to haul down this steep grade.

We ran into two seemingly unimportant pieces of information in this research that may lead to a solution of this problem. Vern Maltais wrote in one of his letters to the author that he seemed to recall Barney telling him, "The crashed craft was shipped out somewhere by rail."[1] Then another piece of information came from UFO investigator Stanton Friedman, who was interviewing in the area prior to the 1980 *Roswell Incident* publication. Friedman was talking to a Datil, New Mexico, postmistress who was there in 1947 and she recalled "a truck with a huge, tarp-covered load coming through Datil late one night earlier that summer." Datil was a small town at the northwest end of the Plains.

As previously mentioned, an old railroad ran from Magdalena to Socorro. In the early days it shipped zinc, lead, and some silver ore. Sometime after WWI the mine was closed. Before the advent of good roads and adequate trucks, local ranchers drove the cattle on foot and shipped livestock on the railroad from some loading pens near Magdalena. The livestock loading and hauling was especially active during WWII due to gas and tire rationing. The railroad was abandoned in the early 1970s, but was quite good enough in 1947 to get the crashed saucer from Magdalena to Socorro. The old grade can be seen from Highway 60 and has much gentler curves than the highway.

Once out of the mountains, the AT&SF railroad ran north and south along the Rio Grande riverbed's flat country. To the north was Albuquerque and Kirtland Air Force Base. To the south was El Paso and the U.S. Army's Fort Bliss with a connection to nearby White Sands Proving Grounds (today's White Sands Missile Range). Alamogordo Air Base was also near there and its name was changed to Holloman Air Force Base in 1948. In the mid 1970s, after the AT&SF tracks between Magdalena and Socorro were abandoned, some 13 miles of them were acquired by the U.S. Government. They were building the huge Very Large Array (VLA) radio telescope facility between Datil and Magdalena on U.S. Highway 60. NASA was originally involved in the project but the radio telescope facility is now run privately, and scientists from around the world do research there. The tracks were laid out in a big "Y" pattern and were used to move the 27 huge, 82-foot-diameter telescope dishes around. So here is one of ufology's greatest ironies. A UFO crashed on the Plains of San Augustin in 1947 some 25 miles southeast of the VLA site, and 30 years or so later, these world-famous radio telescopes were built to try and see if there is intelligent life in the universe. We hope they find it.

There were other considerations that may have influenced the UFO crash being kept quiet. There is the rancher's desire for privacy. There was, and still is, a concern that outside events could cause unwelcome changes in well-ordered lives. Western people need each other, and one sure way to get along is to mind one's own business and not talk about one's neighbors. This, of course, carries with it the unwritten code that one must behave in such a manner that there is no need to be talked about. This posed a dilemma for Barney. He may have witnessed one of earth's greatest events in modern times, but told Vern Maltais he did not want people to think he was crazy when he told others of his discovery.

Barney may not have seen the original *Roswell Record* article of July 8th about the Roswell crash, but he was sure to have heard about it and to have seen the follow-up articles. I do not know what was printed in the weekly *Socorro Chieftain* of Thursday, July 10th or July 17th as these issues are missing from the archives. We do know that Barney listened to the 9:30 evening news and that he had access to articles in newspapers, including the *Albuquerque Journal*. If he had his Plains experience earlier that month, it must have been with some relief and amusement to see the focus on the Roswell announcement about an incident over a hundred miles east and the subsequent denial.

The Dick party had been on the Plains for eight days when news of the Roswell crash story broke. There are two possible scenarios they may have experienced. First, if they were present at the UFO crash site a week or so earlier, they would be quite anxious to see if anyone they saw there had spoken to the media or if word of their experience had leaked out. Even though the daily *Albuquerque Journal* was delivered in the area (a day or two late), it is doubtful that they saw the Roswell UFO articles until they left the Plains site on the 14th or 15th. Second, if they did not have the UFO experience, they would be as surprised and amused as anyone at the news. If again they did find the UFO crash a week earlier, they would be surprised at the Roswell events and greatly relieved that the Plains event was kept quiet. There was probably some speculation among the party that they had seen the real crash site, and that the Roswell saucer/weather balloon fiasco of July 8th and 9th was a cover-up. It would be 33 years (1980) when the stories of the Roswell and Plains crashes were publicized in the *Roswell Incident*.

How then could this secret have been kept for so long? There are a number of things to consider. Fewer than ten civilians probably saw the Plains crash site, and at least half of them may have been in the group of archaeologists. This party left immediately and, it is assumed, had no contact with the others. This leaves Barney and possibly a few other witnesses, besides the landowner who controlled the roads leading into the back side of the arroyo. The loading pit mentioned previously was on one of these roads. It is possible that two groups of civilians arrived independently at the site. Barnett made no mention of several people that we know were there.

What kind of relationship did the ranchers and the Plains-area people have with the government in 1947? For the most part, people wanted as little government in their lives as possible. However, it was important to be cooperative with the government as most ranchers in the Plains area leased grazing land from the state and federal government through a longstanding system of laws and procedures beginning with the Taylor Grazing Act of 1934. During World War II, the federal government exercised its power by withdrawing thousands of acres of grazing land in New Mexico in Socorro, Lincoln, and Don Ana counties for construction of the White Sands Missile Range. Being neighborly and cooperative with the government if you were in civil service, as Barney was, would also have been prudent under the circumstances.

The archaeologists also had a different set of concerns. Not the least of these would be the continuing of federal monies toward various programs at colleges and universities. Then there was the concern about land ownership. Because the federal government owned much of the land containing southwestern archaeological sites, cooperation with the government was important if they were to continue their work in the future. The long arm of government even extended to archaeological sites on private land. Dick was required to obtain permission from the U.S. Department of the Interior to conduct his Bat Cave dig in 1948, although the land was owned and controlled by the Hubbell Sheep and Cattle Company. Federal land is now controlled by the Bureau of Land Management. There were two other concerns the archaeologists would have had. One is the career factor, the concern for one's future employment in their field. The second is that, in the case of the Dick party, the crash experience would have been shared with relatives and friends who were present at the dig. If they had agreed among themselves to keep quiet (and the evidence suggests this), this would be a great motivation for silence.

## CORROBORATIONS

In Corso's book, *The Day After Roswell*, Col. Corso mentions the crash on the San Augustin Plains. Readers may recall his previous position at the Pentagon, and as advisor to the National Security Council and President Eisenhower. Between 1992 and 1997, when he died, Corso gave several interviews. In his book and interviews, he connects the Roswell event with the Plains crash. When asked in a 1997 radio program and other media occasions if the alien technology he had worked on had come from Roswell, he replied,

**"It was from Roswell, but sometimes I get the idea that some of it may have come from the St. Augustin crash."** [2]

In another videotape interview he stated, **"I was privy to the St. Augustin crash."** [bold by author.] [3]

When asked if this crash could have been located at Socorro, 65 miles away, instead of the Plains of St. Augustin, he said;

**"No, it wasn't Socorro; it was down on St. Augustin."** [4]

Col. Corso, while giving a talk on Art Bell's *Coast to Coast* radio program in 1997, described where his Nike anti-aircraft base was located. He commanded this base in the late '50s. He emphasized the location of his base by indicating that a line drawn from north of Roswell through the Trinity site (first A-bomb test site) goes straight through to the Plains of San Augustin.

The implication here is that the two crashes (at Roswell and on the Plains) were related, although the author's research found no evidence of this. However, there is coincidental speculation by some that the Roswell and Plains craft went down within a day or so of each other. In 1957, Col. Corso was placed in command of an anti-aircraft battalion at Red Canyon missile range (at the north end of the 4,000,000-acre White Sands Missile Range area) near the Trinity site. Trinity, as the readers may recall, was where they first atomic bomb was tested. I had been working for three years on this project when Col. Corso made these San Augustin disclosures. I could not ask for anyone in a higher position to support my research.

Col. Corso's statement on at least four media occasions, that there was a Plains of St. Augustin crash, added a new dimension to another theory or two that had been considered by researchers for years. The first theory was that two UFOs had collided somewhere in the Southwestern sky, with one coming down north of Roswell and the other on the Plains of San Augustin. The second scenario was that one craft had experienced trouble, with part of it coming down and the remains (perhaps an escape pod) landing on the Plains where Barnett found it. Just after the turn of the millennium according to *Nature Journal*, Taiwanese researchers discovered that gigantic lightning jets can stream upward from thunder clouds as much as sixty miles into the atmosphere. These lightning jets have been photographed similar to spidery tree-like formations. Perhaps the propulsion or guidance system of the UFO's ran afoul of this atmospheric phenomena.

Lightning and thunder clouds were reported in a band across central New Mexico on the night of July 1 and 2 of 1947. I tend to go with the first theory of two UFOs going down for whatever reason, as three or four bodies were reported at each site. (Six to eight bodies are too many for a small craft.) Barnett was such a credible witness that there is little doubt about his witnessing the crashed saucer with bodies, on the Plains of San Augustin, in July of 1947. As to the Plains crash theory, we are not dealing here with the word of your average aging colonel. In Corso, we have a highly-reputable senior officer at the Pentagon who had been assigned to work with actual crashed UFO technical material and distribute it to private industry. After his stint at the Red Canyon base, Col. Corso was given the command of a Nike AA missile battalion in Germany, as well as other commands.

We also have as a very good witness retired full Col. William Leed who had command responsibilities in the Signal Corps in Korea.[5] Both high-ranking officers, trusted by top generals to lead combat-ready operations in case of hostilities, were placed in extremely sensitive positions overseas during the Cold War. I doubt that either officer was the least bit delusional or prone to cherish fiction in their personal lives before, during, or after military service.

There are two very similar descriptions of aliens from the early July 1947 time frame: Barney Barnett's from the Plains of San Augustin crash, and Col. Corso's (then a major) a few days later, at Fort Riley, Kansas. Col. Corso also had access to secret documents while at the Pentagon in the early 1960s. Among these documents was an autopsy report of alien anatomy from Walter Reed Army Hospital. The author(s) of the special operations manual SOM1-01, if authentic, no doubt had access to restricted material about the aliens' anatomy, their craft, etc. Alien descriptions even from credible sources often differ, especially the size of the eyes. The small alien, EBEs and clones, (lab created entities) as reported, differ little from what Barney reported. These differences would make sense, however, as the developing alien technology probably improved with later EBEs, especially the large eyes. Col. Corso probably alluded to this in a 1997 Art Bell show interview in which he said, "If there are aliens now, I'd like to see them. I don't know where they are. I'd like to see if they compare with the ones I saw."

Various Roswell witnesses (some have been discredited) said they saw bodies identical to those that Barnett described. We must note here that all Roswell witnesses who came forward after 1980, had the opportunity to read Barney's alien descriptions prior to coming forward after the Berlitz and Moore, *Roswell Incident* was published in 1980. The only other mention of gray aliens we could find prior to 1980 was in the book *Interrupted Journey*, 1966. The author John G. Fuller tells the strange story of missing time and the alien abduction of Betty and Barney Hill. Fuller is the first author we could find who mentions gray aliens with large eyes. Barnett mentions "eyes oddly spaced," but does not mention them being large.

Before Barney left the crash site, the military no doubt took his name and address, plus those of other witnesses, and warned them not to talk, before sending them on their way. I might also suggest here that the archaeologists (probably the Dick party) were no doubt checked on at least occasionally over the years. I believe that, as Barney became older and nearer to the end of his life (he was 55 in 1947), he was less prone to intimidation, a sort of "What could they do to me now?" syndrome. As Barney aged and the world approached the age of space travel, he probably told a few more trusted people about his experience. This may explain why he was visited by government agents. These visits may have been precipitated by his retirement in 1959.

Eight months after Barney's death in 1969, Ruth sold the house in Socorro and went to live with her niece Alice in Dalhart, Texas. She remained there until her own death in 1976. She is buried beside Barney, her husband of 52 years, in Dalhart's Memorial Park Cemetery. To be objective concerning the Plains crash story, we must look not only at Barney Barnett's motivation for telling a few friends, but also the circumstances under which they repeated what he had said. Neither Barney nor his friends had anything to gain by telling this story. As a matter of fact, it did not surface publicly until 12 years after Barney's death. The only known public talk Barney gave (except for his valedictorian talk at Humbolt College in 1914) was to his Socorro Rotary Club in 1948, and he made no mention of the Plains crash at that time.

In many cases, the author interviewed Barney's friends and acquaintances, usually several times. As a courtesy, western people will always say such things as, "He didn't talk much about it," "He never said a lot," etc. On second and third meetings with witnesses in an informal setting, many will remember other small things. In many of these cases, feelings not previously perceived come forth. Some even expressed some regret, as in Alice Knight's comment, "I'm just weary of it all," regarding over 20 years of phone calls from all over the world. She said "I'm not sure I want to talk about it anymore." Vern Maltais, in one interview, said that he felt responsible for "dragging Barney's good name down." Beth Danley, Fleck

*Figure 16.2. **The disturbed rock cairn/grave**. Found in the buck brush, by itself, about 150 feet from the gap, these rocks apparently were once on top of some kind of grave. Is it possible that the clean-up crew buried alien mutilated remains here that they could not immediately transport? The cross in Chapter 3 may have originally been here. (The white board at top center was used by the author to orient his compass directions.)*

Danleys' wife, said, "At first we were curious who would travel all this way or call about this flying thing crash, but as time went on, the phone calls got very wearisome." What factors but Barney's honesty and sincerity would compel Barney's friends and relatives to tell the same lifelong story without variance amid much inconvenience, with nothing to gain? Their trust and faith in Barney not only adds to the strength of his story, but also tells something of how those close to him felt about his integrity and character.

Others thought highly of Barnett also. His boss Fleck Danley of the SSWCD said he was "one of the most honest men I ever knew. I never knew him to lie. Not about anything." Barney was highly regarded by his community. He was a board member of the Socorro Electric Co-op and lifelong member of the Rotary Club. Lee Garner, a former Socorro Mayor said of him, "He was a truthful and trustworthy man." William Leed, an Army officer who visited Barney in 1967 said, "I am satisfied Barney was telling the truth. That's all I wanted to know and left." The event apparently affected Barney personally. Ruth, his wife, told her niece Alice Knight, "After that thing in '47, Barney was not quite himself." 5

## THE DISTURBED ROCK CAIRN/GRAVE

About 150 feet east of the gap, we found what appeared to have been a grave site. Even in a local cemetery a mile or so away, graves are protected by rocks placed on top of the fresh soil. In the poorer rural areas of New Mexico largely where those of Hispanic culture live are cemeteries similar to this, as one would find in Mexico. A simple pine box is often used in the burial. It is covered by dirt from the excavation. To keep predators from digging up the recent graves, large and medium rocks are placed on the grave site. The one we found east of the gap in the buck brush was not a grave but apparently had been one.

The rocks were thrown out to the side in an area five foot by six foot, parallel to where we thought something long had been buried. All rocks had been brought in from as far as 75 yards away. Lichen growth on the top side of the rocks indicated exposure to the elements of from thirty to forty years. It would take an average lichen growth ten years to get started. Observations on lichen growth and patterns have long been used to date geologic events, such as landslides. The author first used this concept to date old wagon and stagecoach roads in the 1970s. No lichen was on the bottom of the rocks, indicating they were clean on all sides when placed on the excavation. Later, I would judge that after two to three years they were thrown off and whatever was buried was taken out. Two different digs in the discarded rock cairn area turned up absolutely nothing. There's an interesting observation here though; we found some rust-colored lichen that our BLM botanist identified as *calopaca holocarpa* that grows on rocks but occasionally chooses hardwood. Some of this lichen was growing on the topsides of the rocks that had been thrown to the side. (See Figure 16.2.) No rocks of any kind, with or without lichen were anywhere near the cross when it was found in the gap. Had the cross featured in Chapter 3

*Figure 16.3. **Typical burial practices** in a cemetery not far away. Rocks were a good protection to keep coyotes and other burrowing animals from digging up rural graves.*

originally been placed over what was buried in the rock cairn grave?

We have pretty well suggested (with some degree of accuracy) where items found fit into the crash scenario. The honeycomb, we think, was part of the power distribution process; the metal foil shards were too delicate to be part of the exterior; the heavy and light metal with the strange brown dots no doubt covered the honeycomb inside and out. Even the Harvard-Dick letter had its place in this scenario. But where did the HDPE artifact (Figure 7.1 in Chapter 7) come from? What might its source have been? We know Barnett's description of the bodies. In the Berlitz and Moore *Roswell Incident*, we know Maltais quoted him as saying, "I could see there were bodies inside and outside of the vehicle. The ones outside had been tossed out by the impact." Were there one or more bodies that were mutilated still in the craft? Was the artifact torn or blown out somehow on impact? We know that the artifact was a single organ, and it must have been surrounded by other organs and/or tissue. What state might some of the bodies have been in? Was their gray clothing torn or saturated with body fluid? Was there an odor? Would the remains make the five- or six-hour trip back to the base of the military crew's origin in a closed canvas-top truck?

We need to consider several things. The military might have been prepared for almost any contingency, but probably had not considered the problems of one or more bodies in the back of a truck on a warm day. Although some of the men were no doubt left to begin clean-up. Were they responsible for burying the mutilated remains? Where would the men who came in the truck ride? With no ice or appropriate containers for corpses or dismembered bodies, would it be better to bury the material there? Block ice in those days was at least a two-hour drive east to Socorro, or a one-hour drive south to Reserve, depending which way the convoy turned onto New Mexico Highway 12. To make ice in a modern plant requires a lot of electricity, and it was not until the late fifties that power came into the area from Datil to the north. Some electricity was made here and there by generators connected to windmills. This allowed a light or two for ranchers who had wind generators. Light bulbs were also used to good effect in chicken incubators.

What to do? The only practical thing to get a mutilated body or parts out of sight if you couldn't take them with you was to bury them, and then dig them up later. The soft tissue would no doubt be taken care of by the usual ants, insects, and rodents. It would be too obvious to bury it in the gap where the UFO came in, but much less obvious to bury it. Some feet away in the buck brush, out of sight, which is where we found the disturbed rock cairn/grave. For reasons elsewhere described in this book, the site was monitored and this grave site was disinterred later. It was on private land out of sight. Once the alien remains had been buried awhile, what was left could be taken to the proper place, like the military base of origin or perhaps a military controlled cemetery where it would not be disturbed. To disinter these remains would require moving the rocks off of the cairn to the side, which is exactly where we found them 50 years later. Lichen grows on top of the rocks and none on the bottom next to the ground indicates the cairn had limited use. We believe the cross may have originated there, but the base was not long enough to have been inserted in the ground. After it was made, it probably was laid on top of the rock cairn. It is believed this might have been when the cross picked up some of the rust-colored, slow-growing lichen described earlier. Of course, there is some guesswork here, but the scenario is logical and does make some sense considering what was found at the crash site.

We found evidence under the cross we found in the gap of one or more small fires, brown beer bottle glass, and bottle caps, etc. We believe that the original monitors of the site at first came to check on the rock cairn/grave until whatever was buried there was disinterred. We found out from a previous landowner that the upper arroyo had undergone reclamation and land leveling as evidenced by a berm of dirt on the brow of a nearby hillside probably done in the early 1980s. There is an irony here that this land leveling may have uncovered what the army and some civilian helpers had tried to cover over. Consequently, the weather and rain helped clean the artifacts off, and some were on top of the ground where our 2012 dig crew found them. There were probably one or two men at first visiting the site, and while they were there they picked up other small debris, which had no doubt showed on the surface after rain and wind. We know from the washout of the bridge approach three weeks or so after the crash that a flash flood could undo efforts of the clean-up crew.

We know there were a number of visits over a period of time, possibly from 1947 into the early '70s or so. In the pit we found brown glass and bottle caps and the remains of Spam and sardine cans — this is a fairly typical western man's lunch. Over time as life went on, this occasional visit may have become somewhat nostalgic, and there are signs that family members may have come also on one or more occasions. This might account for the larger fire ring of rocks (found in 2004) near the gap about 25 feet away from the cross/pit area. What did those who sat around this fire know? Who were they? Was the cross later moved over to make a crude, on-the-spot memorial to "the little people" who died there, beings that Barney Barnett had described to his friends and relatives. As far as we know, this was one of the first accurate descriptions, by civilians of the alien beings known as Grays. The large-eyed Grays seemed to be a later innovation.

If the materials found in the arroyo came from extraterrestrials, many aspects of their lives are very similar to ours. We know cellulose material, including cotton, was used in shoe construction, their material was woven like ours and area Chiricahua Indians before us. Also in the shoe sole was found dye, nylon thread, natural and synthetic rubber and a number of things we know and use, although not necessarily in shoe construction. We found at least fifteen or sixteen completely different types of aluminum foil shards of various thicknesses with a strange isotope content. We use aluminum aircraft exteriors and I-beams in various kinds of construction.

The HDPE artifact was the result of a petro-chemical industry, which thrives in our world. The wax was a vegetable product and very similar to our carnauba palm, found in southern climates. Some 12 different elements found in a variety of materials, also appear on our periodic chart of elements. I find it somewhat comforting that possibly on other planets somewhere out there, familiar plants grow and palm trees rustle in the breeze. One other thing a little less reassuring, is that Murphy's Law is alive and well in the universe. If anything can go wrong, it will. It certainly did one night in early July of 1947.

## THE LAST MOMENTS

One of the more interesting finds at the crash site was a variety of moss, *Dicranum flagellare*, on some small and large honeycomb pieces (see Figure 5.4). In researching the moss, we were surprised to find that it did not grow in New Mexico. It is very common in 30 states in the eastern United States, but less common in western states. It does grow in three states bordering New Mexico — Arizona, Colorado, and a tiny piece of Oklahoma. *Dicranum* moss, as it is generally called grows in a variety of habitats. It is found in low forested areas on rotting logs and soil along stream banks. It is also found in higher areas, on rock outcroppings where moisture is retained. It is relatively common in pine forests and meadow habitats.

How the craft became disabled is open to question. But we do think it was forced initially into a descending trajectory by a power failure. The craft may have had some lateral movement and was able to make a turn into its final descent some miles north of the arroyo. Evidence found at the crash site indicates there were perhaps at least three or more contacts with the earth before it finally reached the Plains. The craft apparently grazed some high ground, losing some of its external skin. This exposed some of the honeycomb. Subsequent skip-downs may have picked up some *Dicranum* moss. In Figure 5.4, we see a pine needle embedded in some of this moss in a large piece of honeycomb. The pine needle could have come from brushing a pine tree north of the Plains, but the craft must have picked up the moss 200 to 300 miles away because pine trees are not endemic to the Plains. The craft trajectory indicates it was coming from the northeast. Three hundred miles in that direction is Colorado. Is it possible that the Plains UFO skipped its way like a flat stone on water for several hundred miles, picking up moss and brushing pine trees as it lost power and altitude?

We surmise from the wreckage in the upper arroyo that there was another touch-down, leaving debris that the dig crews found in 2011 and 2012. The next and final contact with the ground we believe occurred in the lower arroyo where it hit much harder, resulting in a vertical or widening split by an explosion or decompression. This area, about 20 yards in diameter, yielded some 15 metal foil shards, the tiny shoe sole, some wax pieces, and the HDPE artifact. Indications are that the artifact had previously been torn from the body that contained it and was subjected to "high heat or explosion" before exiting the craft. Before the downed craft came to a stop, it plowed over a 100-foot-long "shallow" trench that Gerald Anderson called a "furrow" when he described it to Kevin Randle in 1990. The buck brush did not grow back here after the military leveling equipment disturbed the roots. This was later known to our researchers as "the gap."

Our story did not begin until about 10:30 or 11:00 a.m. the next morning when Barney Barnett, traveling west on N.M. Highway 12 looked over and saw the sun glinting on a metal surface. He told Vern Maltais, "I was working with the rancher who lived near there and knew he was not building anything...I thought it might be a downed aircraft and went over to investigate." Soon two other civilian groups had arrived and within the hour the military arrived also. The rest is history. There was only one person to our knowledge who we interviewed that had learned about the crash from other than a New Mexico source. As mentioned before, this was retired Army Colonel Bill Leed.

It is doubtful that Barnett came up on the scene where two or three bodies were laying peacefully near the crashed, split-open craft. There may have been debris including bodies whole or mutilated strewn for 200 feet or more. This would not be the coloring book scene depicted by some UFO artists of neat whole bodies laying in a row near a craft with a gash in the side.

As I look back at early UFO writers and artists, I am amused by archaeologists depicted with scout hats, walking sticks, and knee socks casually looking at a scene reminiscent of a "fender bender" in a Sears and Roebuck parking lot. There were two indicators of the horrific scene that I didn't understand for years. The first was from Vern Maltais. He said Barnett told him, "It was horrible; I can't even describe it, don't want to." My recent thinking has shifted somewhat to include as a logical assumption that if debris was scattered in a long swath, why not body parts? The second indicator was the rock cairn/grave site found down in some buck brush, which was discussed earlier in this chapter.

Sooner or later in this manuscript, we must face the possibility that one or more of the aliens in the ship that crashed in the arroyo was a clone (artificially created biological being apparently in a gray alien-looking form). Col. Corso, the Pentagon insider, alluded to artificial parts being added to cloned EBEs (Extraterrestrial Biological Entity). We believe definitions are needed here concerning alien clones and alien androids. The clone is an identical copy of a living organism, usually created in a laboratory environment by a non-sexual process. Sometimes the word android is used when speaking of aliens. An android is a robot, as we know it, or synthetic organism, usually designed with mechanical parts. We do not believe android is applicable here.

The artifact, as seen in Chapters 7 and 12, we know is made of a special kind of high density polyethylene (HDPE). It seems to have folds of material also of HDPE, which probably connected it to parts of whatever organism it came from. We have reason to believe it may be the central part of an alien cloned heart-lung artificial organ. As far as we know, this is the only artificial body part connected to a UFO crash in civilian hands. We are sure the government may have similar or other artificial parts from other alien crashes. This part was probably missed by the clean-up crew because it was covered with sandy loam soil and may have been buried deeper when equipment was used to level the site. We found it on the surface; it probably took many years of wind, rain, and erosion to uncover it.

Other scientists looked at the HDPE artifact. One wrote, "Mr. Campbell, we are intrigued by your project and the thoroughness of your research. According to the samples you sent us, we concur that they are HDPE and we also agree with the Braun Intertec analysis that the object was amorphous." [6] We asked the research facility to give us a definition of HDPE a layman would understand. They wrote, "High density polyethylene is a strong, chemically resistant plastic. The flexibility of HDPE depends on its degree of crystallinity or degree of ordering its molecular chains and is most flexible when it is amorphous (no crystallinity or molecular order)." As mentioned before, the author was on the History Channel *UFO Hunter* television program in 2008. The episode was called "The Real Roswell" and a segment was featured on the HDPE artifact. Independent lab tests again showed it was high density polyethylene.

I also asked the above lab if they could identify the artifact's possible use, and would the material be consistent for artificial body organs or perhaps an artificial heart? We also, at their request, sent photos of the artifact (including the one in Figure 12.9), showing the broken edge. They wrote, "Thin sheets of amorphous HDPE should be sufficiently flexible and suitable to be used in the walls of important body organs, including those of an artificial heart." They also said it was possible to introduce other alterations in chemical structure that would increase the degree of flexibility of HDPE.[7]

Up until this letter was received, we had assumed that the lack of crystallinity was due to the heat or hot environment of one or more internal explosions when the craft crashed. If the lack of crystallinity was due to its chemical makeup in the "ordering of its molecular chains," the lab suggested that the HDPE amorphous artifact might be extremely valuable to medical science.

One or more explosions on the ground may have thrown out the already exposed body part and gave it a quick exposure to heat, which may have caused the slight melting. Our original find of the HDPE artifact may be able to tell us a little. (See Chapter 7.) The chemist who originally analyzed the artifact said in the fourth paragraph of his report, "The polymer has undergone a transition to a more random state (perhaps by heat or chemical treatment)." Other scientists all thought that it had been in an explosion or very hot environment. Polymer starts to melt at 248° to 266°F. (120–130°C). So we know at least this temperature was reached in another explosion that occurred when the artifact may have been blown out of the craft. We believe on one of these arroyo explosions, the all important foil shards and the small shoe sole also exited the craft. The shards, which we think may have been sandwiched between the inner and outer upper coverings of the craft, blew out when the outer surface was ruptured. Partial melting on the artifact can be seen in the dark areas in Figure 12.4. A sand texture was found on one side of the artifact indicating it may have been hot when it hit the arroyo surface.

By the 50th anniversary of the Roswell event in 1997, commercialism and "hype" more or less replaced good research and objectivity. While Roswell crash theories and questionable witnesses were considered, critiqued, and discarded over the years, Barney's simple story no longer fit the researchers' complicated "Roswell" local event scenarios. Charles Zigler takes a look at the Roswell/Barnett scenario in the context of versions of a modern myth. In version one, he has Barnett and the archaeologists, correctly, stumbling upon a crashed UFO on the Plains of San Augustin.

Then the myth undergoes a three-step change. By version three, Barnett and the archaeologists find the downed craft north of Roswell on the Foster Ranch. Another variation of the myth has Barnett traveling as a mentor with a group of archaeology students and finding a downed craft, further north of Roswell. This version is in some of the 1990s UFO books, but has been proven to be totally in error. Zigler has Barnett eliminated by version five, and a new group of archaeologists are introduced. This would be the well known Texas Tech southwest anthropologist, William Curry Holden. After trying to "shoehorn" Holden into the Roswell scenario mold, he was wisely eliminated. In Zigler's version six, all references to Barnett and any archaeologists are eliminated. Instead of thoroughly investigating Barney and his claim, early UFO researchers tried to put him and his compelling story in places he had not been and connect him with events he knew nothing of. Like a cheap detective novel, where a character is introduced early and is not part of the final solution, he needs to be written off. To add insult to injury, we find the final attempt to eliminate Barnett in a popular UFO magazine.

**WRITING OFF BARNEY BARNETT**
In the summer of 2003, an article appeared in the *IUR* (*International UFO Reporter*) entitled, "Barney Barnett's Crashed Saucer: Where Did It Come From?" [8] The article was written by two respected Roswell writers, one of whom passed away in 2006. The suggestion of this article was that Barney was influenced by popular culture magazines in the late 1940s and early 1950s that "could have fueled his imagination." Two articles and one book cited in the *IUR* article were written by Frank Scully. The articles cited that were written by Scully appeared in *Weekly Variety* magazine in the fall of 1949 and *Time* magazine in 1950, along with articles by others appearing in two earlier publications, *Reader's Digest* and *Pageant*. Scully's new book in 1950, entitled *Behind the Flying Saucers,* was also referenced. None of these articles appeared before the fall of 1949. One has to ask how could Barney's imagination be fueled in 1947 by articles, magazines, and a book that would not be published for two and a half years? In all fairness to the authors of the *IUR* article, they apparently made their conclusions based on the assumption that Vern Maltais was the first to hear about the crash in February of 1950.

Information received from two witnesses indicates that

Barnett told his boss, Fleck Danley about the crash "in early summer of 1947." [9] Alice Knight, Ruth's niece recalls hearing the account at Thanksgiving of that same year when they visited the Barnetts in Socorro. Barney and Ruth never to our knowledge wrote about the event in those days before Barney retired because of job security and credibility in his work. There is some evidence that pressure was put on Barney to remain quiet after a meeting between Fleck Danley and USDA representatives in Albuquerque. Military seizure of land for the White Sands Proving Grounds may have been a consideration in those days.

There was a suggestion that barber shop literature was an available source. That is possible after 1948 but it would not apply to Barney in any case. And another factor that is relevant here is that he was completely bald in those years. Alice Knight, Ruth's niece, said he was somewhat sensitive about it and usually wore a hat. Page 59 shows him circa 1922 with a very receding hairline, and another photo in 1948, page 55, has him with a hat standing with Ruth in their front yard. Alice said that for years she had trimmed what hair there was until it was all gone.

In 1947, one of the authors of the original *IUR* article was four and a half years old, and the other was not born until 1949. two and a half years later than Barnett's 1947 discovery of the arroyo crash site. In any case, small town barber shops seldom had anything but very old magazines and newspapers, and a comic book or two. There might be an *Esquire* magazine put aside for mature customers. The author wrote a letter to the *IUR* editor in 2003, which was published with Barney's circa 1922 photo. In the next issue, this author is quoted, "Barney's Plains story stands on its own. He was a quiet, unassuming man, certainly not prone to cherish fiction and pass it on to his friends. His story for those who care to study it is sincere, consistent, and lifelong." [10] The Barnett story remains amazingly consistent. While some of the main Roswell witnesses, once touted as the core of the Roswell story, have come and gone, and their accounts must be considered at best unreliable. All aspects of Barnett's story that could be researched are in these pages. The reader can be the judge.

Gerald Anderson may have had more support from the UFO community than he knew in the early 1990s. John Carpenter, a psychiatric social worker (photo with Gerald in Figure 5.1) wrote an article for the Mutual UFO Journal (MUFON) in 1991. In the conclusion of his article, he gave UFO researchers some good advice. He said, regarding Kevin Randle and Don Schmitt's rejection of Gerald's story, "Premature conclusions and hasty public debate reflect a war of egos rather than the work of researchers with a scientific approach. [11] It could not be better said. Carpenter was one of the few in those days who could see the rocky road ahead that many UFO researchers would travel in their "war of egos."

Five members of the Anderson family had over a year to discuss, solidify, and mull over what had probably been one of the most important events of their lives. It was the fall of 1948 that Glenn Anderson took his job at the high security Sandia Corporation. This was more than enough time to generally agree on what had transpired on that warm July morning of 1947. This was not a story conjured up and agreed on for someone else. In their wildest dreams, none of them ever expected to tell of the event, outside of the family, but it did occasionally happen. Early in the 1990s Stanton Friedman talked to a man from the Anderson's church who had heard about the crash from Gerald's dad, Glenn. This indicates that the Anderson adults were telling some people about the UFO crash. This was probably before Glenn was interviewed at the Sandia Corporation and the security check had begun. We understand Linda Moulton Howe also interviewed the man and he related to her basically the same information.

The New Mexico newspapers occasionally related something of interest to UFO investigators. Chuck Wade's father in Corona had known Mack Brazel, who originally found the Roswell UFO wreckage. Some witnesses from those days knew Brazel, who came to Roswell occasionally for Foster Ranch business. In the second week of July 1947, during the initial Roswell newspaper and radio coverage of the saucer/weather balloon stories, it was said that Brazel was "well escorted" by the military around town. Loretta Proctor, who knew him, said in several interviews that apparently Brazel had "come into a little money." Brazel was a ranch manager/wrangler of very modest means. Loretta said she had heard that Brazel had bought into a meat storage locker business near Alamagordo. He was also reported to have been driving a new pickup truck around town. Shortly after this, he was not seen in Roswell and left Foster Ranch. The author ran into a small, brief boxed ad in a 1949 Otero County Times Newspaper. The ad read, "**NOTICE:** We are not accepting any more processing at **Otero County Locker Plant – Closing Out –** Locker room will be closed as of JULY 1, 1951. **W.W. BRAZEL.**" Was this ad the last piece of the Roswell story puzzel? It may be.

Earlier in this research we contacted a granddaughter of Mack Brazel, who lives in Oregon. She is Chere Winters. Mack and his wife had five children, two daughters an three sons. Chere is the daughter of Paul Brazel, the eldest son. Chere said her grandfather did have a meat packing and storage locker business near Tularosa on Hwy 54 in 1949, the year she was born. Her father, Paul, worked there also. The Business was sold in 1951 three years after she was born. She further explained that in 1953 the family moved to Oregon, but Mack and his wife returned to southern New Mexico a few months later.

As an adult, Chere raised a family and eventually took over the Brazel holdings in Oregon. She established a very nice cranberry-growing operation on the scenic Oregon coast. She hopes someday to write about her grandfather's life and his role in finding the Roswell crash debris. Mack Brazel was a Mason and he died in 1963. He now rests beside Maggie, his wife of over 45 years in the Tularosa, New Mexico Fairview Cemetery.

**REFLECTIONS**

Time on the Plains is not measured by urban standards. The daily lives of the people are totally dictated by the needs of the livestock and the land, between daylight and dusk, and from season to season, year after year. These are sturdy people distilled from the Depression survivors and the hardy pioneer. Although they are very much aware of their roots, their focus is primarily in the present rather than the past. Those with whom I have spoken in the area do not necessarily believe or disbelieve in a Plains UFO crash; it is just not relevant to their low-key, well-ordered western lifestyle.

As this book was in its final phase, I visited the site above the aspen grove once again. On this day I was feeling nostalgic. As I gazed out over the cloud-tufted sky and followed the shadows across the landscape, I recalled the day I had first picked up a handful of soil and let it trickle slowly to the ground. That day I stood in the arroyo and looked back at the mountain where I now stood and realized that the arroyo had once been a part of it. Each year in the arroyo the spring winds blew back some of the soil, with a little more being washed into the Plains when the rains came. Even the yearly freezing and thawing helped bring lighter material to the surface; through constant weathering, the inexorable forces of erosion move the loose soil and sand down the mountain through the arroyo and onto the Plains. Over time, as the water courses shift and the wind blows, the arroyo gradually revealed its secrets.

From my vantage point high on the mountain, the evidence was overwhelming. Something had happened down there. We certainly did not have all the answers, but we had some intriguing pieces. Had the crash site that Barney, Col. Corso, retired Col. Leed, and Vern Maltais referred to been relocated? Yes, we believe it had. Was this the "surveyor's" crash site Jesse Marcel mentioned in 1978 to Bob Pratt? The 60 to 80 miles west of Carrizozo certainly fits our site description. Although the Plains are fifty-nine miles long and five to fifteen miles wide, there is only one highway, NM 12, that runs the length of the west side. If it was improbable for Barney to be at the Roswell site, then he sighted his crashed UFO down there off the Plains road. The archaeologists according to the Harvard letter were camped a few miles away and probably met Barnett there.

When we considered the fifty or so items we found in the arroyo, we began to realize that there was something here we had not seen in other crash scenarios, including Roswell. We had a physical trail of artifacts from the arroyo sands to the laboratory, and then to public dissemination. Total possession of something from its earliest history is known as provenance. We certainly had this, but proving that it was not from our time and place required something greater. As far as I know our project was a civilian first when we found part of the UFO propulsion system.

It took us a while to realize that the brown dots were not a burnt residue at the end of each honeycomb channel. It was a revelation to find that they were an integral part of the UFO power system using dysprosium and terbium in what Steve Colbern labeled "inverse gravity propulsion."

We have found many answers, some quite unexpected, but still have many questions. The gold and copper microscopic wires in the HDPE artifact carried current to what? What material flowed through this artifact that left these wires visible in the starch residue? What was the source and power of the energy that flowed through the honeycomb piece? And was it sandwiched between the heavier and lighter metal shards? This supposition was borne out on the 2012 dig when a piece of the lighter skin (W-1103) was found still attached to the second honeycomb. Was it an integral part of the functioning UFO, as the identical honeycomb wreckage from the 1971 Japanese UFO crash indicated? And what were the circumstances of the apparently aged and semi-malfunctioning UFO making one final turn to crash on the Plains where Barnett found it? How did the zinc wafer function? In the shoe sole research, we learned that there were nearly 300 known width and length combinations on earth. Our little shoe sole fits none of these classifications. I had one last question that still haunts me. Who or what had worn that tiny shoe?

As a cool wind came up, I looked out over the Plains again and pictured a wind-rippled, shimmering lake and glaciers receding into the high mountain valleys . . . I was brought back to reality when I saw down on the Plains a tiny speck, a ranch pickup trailing miles of dust. In the distance, I could see dust devils here and there spiraling up from the loose, caliche soil of the playas, almost white against the blue-gray haze of the distant mountains. As I gazed out over the broad prairie landscape where the sky met the desert, I realized that time and circumstances had rendered the complete story of the San Augustin crash an enigma.

For a while, there were some who knew parts of the story, but almost all have come and gone and faded into obscurity. This is why this story had to be told. Down in the dry, scarred, windswept arroyo can be found the discarded items of yesteryear depicting life's hardships, and some of life's simple pleasures amid some strange unexplained materials from someplace else. If we listen carefully, the wind also speaks of what may have been one of the world's most significant events. From the mountaintop, I viewed a distant cloud momentarily lit by lightning followed by a distant echo of thunder. The lightning reminded me of a window barely opening and then closing again . . . like the Plains story. As the wind died down and the thunder rolled off, stillness came over the landscape, and then all that could be heard was the sound of silence.

## CHAPTER 16
## REFERENCES

1     Campbell, Art. Correspondence and communication with Vern Maltais, 1998-2001.
2     Corso, Col. Philip J. (Ret.), audiotape interview, Art Bell radio program #970706D, Chancellor Broadcasting Co., 1997.
3     Corso, Co. Philip J. (Ret.), videotape interview with Ted Loman, *Beyond The Roswell Files*, UFO Collector Series, Vol. 188, 1997.
4     Knapp, George, "The Untold Stories of Col. Philip Corso," videotape interview, UFO Collector Series, Vol. 318, International UFO Congress, Inc., Westminster, Colorado.
5     Bragalia, Anthony. "The Other Roswell Crash," http://ufocon.bogspot.com/2010/05/other-roswell-crash-secret-of-plains.htm, May 2010.
6     Campbell, Art. Correspondence and communication with scientists concerning the artifact.
7     Ibid.
8     Randle, Kevin and Carl Pflock. "Barney Barnett's Crashed Saucer Where Did It Come From?" *IUR*, Spring 2003, pp 15-18, 24-25.
9     Ibid. Danley, Fleck.
10     Campbell, Art. *IUR*, Summer 2003, p. 32.
11     Carpenter, John S. "Gerald Anderson, Truth or Fiction?" *Mutual UFO Journal*, no. 281, Sept. 1991.

### OTHER MATERIALS USED & CONSULTED

Road Map of New Mexico, 1947, New Mexico State Highway Department, State Highway Commission, Santa Fe, New Mexico.

*Alamogordo News*, July 10, 1947, C. W. Morgan &Sons, publisher, Alamogordo, New Mexico.

*Albuquerque Journal*, July 19, 1947, Journal Publishing Co., T.M. Pepperday, publisher, Albuquerque, New Mexico.

# Index

## A

Air Materials Command Center, Wright Patterson, 69
Aerial Phenomena Investigations (API), 120–21
Ake, Marvin and Trudy, 120
Alamogordo Air Base, 45, 68, 119, 126
Alamogordo, New Mexico, 45, 63
*Albuquerque Journal*, 127
Albuquerque, New Mexico, 28, 66, 69, 77
alien
    autopsy, 115
    bodies, 106–107, 133
    clothing, 107
    description, 67, 74, 99, 106–07, 130
    eyes, 74
    Grays, 107
    honeycomb (Pinkney), **35**
    implants, 113
    injuries, 107
    little people, 55, 60
    origin, 25, 41, 53, 98
alluvial fan, 4, 16
*Apollo 13*, 37
archaeologists, 28, 66, 73, 78, 88, 89
Aluminum Association, 17
Amarillo, Texas, 57
American Antiquity, 76–77, 85
Amijo, Gilbert, 106, **119**, 122
Anderson, Gerald, **27**–30, 34, 39, 41–42, 79, 88, **124**, 135
Anderson, Glenn, **124**, 135
Army Air Corps Intelligence, 75, 116
arroyo, 3, 16, 112
artifact, **49**, **51**–52, **92**–**93**, **95**, **97**, 134
artificial body parts, 53
AT-6C Trainer crash, 120
Aztec, New Mexico, 36
Aztec UFO Conference, 18, 28, 41
Aztec UFO Crash, 95

## B

Baca, Harold, 55–**56**, 60, 71
Banks, Tom and Nancy, 41
Barnett, Landon (Barney) Grady and Ruth, 37, **55**, **57**– **60**, 72–73, 75, 77–79, 84, 86, 88, 106, 129–130
Bat Cave, 1, 65, 75–76, 81–**83**, 87, 129
*Behind the Flying Saucers*, 134
Belfort, France, 58–59
Bell, Art (*Coast to Coast* Radio), 129
Berliner, Don, 27–28
Biefeld-Brown, 44
Bigelow, Robert, **27**–29
blowouts, 85
body part, 28
Boise City, Oklahoma, 57, 59
bowl-shaped grid, 40
Braun Intertec Northwest, 91

Brazel, Mac, 34–35
Brew, Dr. J.O., 75, 78–79, 84, 87
brown dots, 35, 37, 45–**46**, 47
Brown, Harold, 77, 78
Bureau of Land Management (BLM), 32
Bursom, Jr., Holm, 60
buck brush, (*coleogyne ramosissima*), 7, 9, 42

## C

calcium carbonate, 52
Cambridge, Mass., 84
Camp Owen Biernie, 58
Campbell, Art, **20**, 28
Calvary, 15th, 58
Camp Dodge, Iowa, 59
carbonized fiber, 53
Carey, Tom, 75
Carkin, Renee, 40–41, **46,** 135
Carpenter, John, **27**–28, 133, 135
Carrizozo, New Mexico, 119, 120
Carter, Gordon, 77
Cavitt, Sheridan, 69
cellulose plant fibers, 52
cellulose starch, 94–96
Center for UFO Studies (CUFOS), 105
ceramic duck, 93
ceramic material, 54
*Challenger,* 37
Chiricahua Indians, 10, 83
Chicago Natural History Museum, 85–86
chickens, 122
Chaco Canyon Conference, 78–79
charcoal, 54
Cimmaron County, Oklahoma, 59
circle/triangle symbol, 40
civil engineer, 57
Cleveland, Agnes Morley, 84
cloth fiber, 102
C-N Basin, 81
Colbern, Steven, 16, 22, 121, 124
    Colbern, Metal Sample Conclusions, 38
    Colbern Report, 23–25
    Colbern Summary, I-beam and brown dots, 47
Colorado Museum of Natural History, 59
Columbia, 37
Continental Divide, 2
copper wire, 96
Corona, New Mexico, 19
Corso, Col. Philip J., 97–99, **113**–117, 129–130, 133
*Crash at Corona,* 42
crop circles, 41
cross, **13, 14**
crystalline structure, 113

## D

Dalhart, Texas, 59, 124, 130
Danley, Beth, **64**, 69, 88

Danley, Fleck, **63**, 68, 70, 124, 131
Danson, Edward, 84
Datil, New Mexico, 13, 15, 50–51, 64, 69, 84
deer mice (peromysas maniculatus), 31–32
Demb, Sarah, 76
deuterium gas, 46
Dick, Herbert W., 68, **75,** 77–78, 80, 82, 88–89, 129
Dick, Martha, 77, 86–88
*Dicranum flagellare* moss, 133–34
dig crew, 2012, 41
dig crew, 2003, 10
Division, 88th, 58–59
Drake, Joe, 119
dysprosium, (DY), 45, 136

## E

Eagle Guest Ranch, 19, 49
eclipse, 41
Ehime Prefecture, Japan, 36
Eisenhower, President Dwight D., 114
El Palacio, 85
equipment timelines, 26
exotic coatings, 73
explosion zone, 53
extraterrestrial, 25, 38, 47
Extraterrestrial Biological Entities (EBE), 53, 109, 115

## F

Fay, Cara, **29**, 124
fabric scraps, **53**
Feinstein, Seth and "Annie," 43
flash flood, 7, 13, 33
foil shards, **18–19,** 36–37
Fort Bliss, Texas, 128
fosgene gas, 58
Foster, Jodie, 2
Foster Ranch, 69
French trenches, 58
Friedman, Stanton T., **27**–29, 35, 39, 60, 66, 73, 75, 79, 88, 120
FT-IR Spectroscopy, 54
Fund for UFO Research (FUFOR), 88

## G

Gallinas Mountains, 42
Gallup, New Mexico, 15, 17, 124
gap, 7–8, 42, 52, 85, 111
Great Basin, 2
Glenn, John, 114
gold wire, **94**
Goodnight Baptist College, Texas, 57
Gravette, Arkansas, 124
gray rubberized fabric, 53
Green River, Wyoming, 123
guidance system, 42

## H

hantavirus, 8
Hartford, Minnesota, 29
H-beam, 40
*Harvard Gazette,* 79
Harvard University Peabody Museum, 67, 75–76, 79, 84, 133
Haut, Lt. Walter, 69
HDPE artifact, 2, 32, 40, 53, 107, 111, 113, 132–34
heavy equipment, 35, 42
helicopter, 42
hieroglyphic writing, 40
high country, 64–66, 70, 71
high density polyethylene (HDPE), 93, 98
Hill, Betty and Barney, 130–31
Hispanic culture, 13, 52
History Channel, *UFO Hunters*, 98, 134
Holden, Dr. William Curry, 132–34
honeycomb artifact, **30**–34, **35, 36,** 39, **41–43, 44–45,** 46, 131
honeycomb, Japanese, 35
Honeycomb Sandwich (drawing), **36**
Horse Springs, 3, 13, 15, 64
Howe, Linda Moulton, 41, 135
Hubbell Cattle Company, 76, 87, 129
Hubbell, James L., 76
Humboldt College, 57
Hurt, Jr., Wesley R., **76,** 78, 85, 124

## I

I-beam, **39, 40, 46**
ICP-MS, 14, 19, 24, 26, 45
Indian artifacts, 59
Indiana University, 85
Infrared Microspectroscopy (IR), 52
*Interrupted Journey,* 128
inverse gravitational propulsion, 44
isotopes, 17, 20, 22, 27, 34, 38, 47, 73
*International UFO Reporter (IUR)*, 134

## J

Japanese, 59–60
J.I. Case Tractor Company, 59
Johnston, Langford, 3
Jun-Ichi, Yaoi, 35

## K

Key, Andy, 24–**127,** 128
Key, Mildred, 125–27
Key, Phoebe Ann, **127,** 129
kangaroo rat (pack rat), 8
Kansas City, 27
Kansas City NICAP, 50
Kimbler, Frank, 42
Kirtland Air Force Base, 122, 128
Knight, Alice and J.B., 56–57, 59–60, 66, 70, 89, 124, 132, 135
Korean War, 114
Kropp, Cynthia, 41

## L

laboratory equipment, 19
Lake San Augustin, 85
land transfer, 127
Las Vegas, 46
Lawrence, Merle, **10**
Lazar, Bob, 37
Leed, Col. William, 55, 70, **71**–73, 112–13, 115, 128–130
Leir, Dr. Roger, 22, 104, 113, 114
LeHarve, France, 59
LeMaster, John, 39–41
lichen, 14, 132
light microscopy, defined, 23
Little People (aliens), 55, 60, 67–68, 72
Liverpool, England, 58
sandy loam soil, 37

## M

MacDonald, David (MUFON Director), 121
Magdalena Mountain, 64
Magdalena, New Mexico, 15, 68, 80, 121
Magenheimer, Lt. Eddie, 120
magnetic balance, 59
magnetic energy, 44
Magnetic Resonance Imaging (MRI), 93
magnetorestrictive alloy, 45
Maltais, Vern and Jean, 55, 59, 66, **67,** 68, 71, 73, 80, 89, 106, 128, 134
Marana AAF, 120
Marcel, Major Jesse, 36, 39–41
Marcel, Linda, 74
McArthur, General Douglas, 114
McCrone Associates, 102
McKnight, Daniel, 76, 85
mechanical energy, 44
metal detector, 30., 43
metal
    large heavy piece, **29, 32,** 38
    light piece, 31, 38
Mexico, 41
Mexican vaqueros, 52
microtechnology, 44
microwave energy, 44
mis-diagnosed I-beam, 121
microwave frequencies, 45
missile types, 122–23
Missouri, 27
Mogollon Culture, 86
monsoon rains, 85
Monte Vista, New Mexico, 119
Moore, L.B., 125
moss, 43, 133
motherlode, 27, 111
MUFON, 120–21
MUFON Star Team Investigators, 120–122
Murphy's Law, 42

## N

Nancy, France, 59
NASA, 123, 129
National Enquirer, 73
National Radio Astronomy Observatory, 2
National Shoe Retailers Association, 101
Native Americans, 51
natural rubber, 52
Navajo Copper Company, 57
Navajo crew, **10,** 20, 26, 29
Naval Reserves, 27
New Horse Springs Store, 120–21
New Mexico Highway 12, 37, 87
New Mexico State University, 19, 27
*New York Herald Tribune*, 85
*New York Times*, 85
Nike Ajax Missile, 123
Nininger, Dr. H.H., 59
Nippon Television, 35
Nixon, Richard E., 95
No Life for a Lady, 84
nonterrestrial, 25
North Island Naval Air Station, 50
nylon thread, 101

## O

Oakland Polytechnic College of Engineering, 57
Oberlin hole, 3
Odessa Crater, 59
Oklahoma Fish and Game Commission, 60
Olsen, Mark, 41
Open Minds Organization, 28, 31, 111

## P

pack rats, 31–32
Paleo-Indian agriculture, 80, 85, 87
Panhandle Radio and Electric Company, 59
Paris, Antonio, 121–122
Park Street, 55
Pearl Harbor, 63
pedorthist, 106
Petersburg, Virginia, 58
Pie Town, New Mexico, 15, 71
Plains of San Augustin, **1, 4, 10,** 75
Plains of San Augustin Controversy, 34, 75
plane crashes, 119–120
playa, 81
Pleistocene ice invasions, 2, 10
Polarized Light Microscopy (PLM), 52
polyacylonitile (PAN), 53
Potter, Paul E., 46–47
prairie dog, 8, 13, 42, 51
Presbyterian Church, 55, 66
prickly pear cactus, 42
Project Skyvault, 44
provenance, 133

**139**

## R

railroad, Atchison, Topeka and Santa Fe, 67, 128
Rao, John, 20
rare earth elements, 45, 47
Randle, Kevin D., 33–34, 75, 88, 125, 135
Ramey, General Roger, 69
rattlesnakes, 7–8, 42
Red Canyon, 128
Reserve, New Mexico, 13, 84–85
Rio Grand River, 63, 66, 69
rock cairn/grave, **131**
rodent activity, 16
Roswell, 39, 44, 73, 79–80, 125
*Roswell Incident* book, 67, 72, 75, 78
Rotary Club, 66, 69
royal flush, 27, 29
rubber coating, 53

## S

St. Lawrence University, 72
Salado Soil and Water Conservation District (SWCD), 35, 55, 60, 63, 65 (map), 79
Sandia Corporation, 124
San Francisco Mountains, 77
Santilli, Ray, 115
San Mateo Mountains, 51
SCS surveyor, 68, 73
Scanning Electron Microscope (SEM), 19, 26
Schmitt, Don R., 75, 125
Scully, Frank, 50, 134
second lieutenant, 58
Sherman County, Texas, 59
Shikoku Island, Japan, 35
Shirley, Benjamin, **10**, 31, 124, 134
Shirley, Dick, **10**
Shogun, TV miniseries, 36
Shroud of Turin, 52, 102
shoe
  bottom, **52, 101–02**
  drawing, **103**
  sole, 28, 52, 67, 101, 116, 133
shoe last, 104
Situation Red: The UFO Siege, 95
Signal Corps U.S. Army, 72
skin (UFO), 34, 37, 43
*Sky Magazine of Cosmic News*, 60
sky path, 41
sky path-eclipse, 40, 41
Smithsonian Institution, 59, 88
*Socorro Chieftain*, 15, 45, 59, 63, 73
Socorro Electric Co-op, 55
Socorro Lions Club, 55
Socorro, New Mexico, 15, 55, 63, 80
Soil Conservation Service, Mosquero, 59
Soule, Capt. Floyd, 119
solidified chicken fat, 51
Steinman, William S., 95
stereomicroscope, 52
starch, **94**, 114
Stratford, Texas, 59
stratographic survey, 87
Stringfield, Leonard H., 97, 107, 109
sulfonated oil, 114
supervising survey crews, 73

## T

T-bar, 14
Taylor Grazing Act, 129
Terbium, 45
Texas Panhandle, 59
*The Day After Roswell*, 105, 114, 129
*The Roswell Incident*, 88
*The Roswell Report: Case Closed*, 122
*The Sky Determines*, 51
*The Sky: Magazine of Cosmic News*, 59
Tilden, Mervin (Murff), **10**, 39, 43, 45, 124
*Time* magazine, 1950, 134
tinsel, 34
toroid, 46
trace element, 38
Trailways Bus, 69
transducers, 44, 47
trash, 33
triangle/circle emblem, 40, 41
Trinity site, 114
Tupperware, 113

## U

UCLA, 22
UFO, 16, 34–37, 55
UFO crash damage, 34–36, 111
*UFO Magazine*, 115
UFO propulsion, 46
University of Michigan, 78
University of New Mexico (UNM), 75, 79
University of Pennsylvania, 66
*Unsolved Mysteries* TV program, 28, 79, 88, 135
Upper Gila Expedition, 77–78, 84
U.S. Army, 58
U.S. Army Air Corps, 35
U.S. Air Force, 35, 119
U.S. Department of Agriculture, 86
  Soil Conservation Service, 59, 69, 70
U.S. Department of the Interior, 127
U.S. Forest Service, 86
U.S. Highway 60, 64, 68, 71, 81, 123
U.S. Navy Seabees, 27
*USS Boxer*, 27
*USS Buford*, 59
*USS Madawski*, 59

## V

VLA (Very Large Array), **2**

## W

Wade, Bobby, **11**
Wade, Chuck (and Nancy), **8,** 10, 14, 15, **17**–18, 23, 27, **29**, 39–42, 111, 124
Wade's Bar, 19
wafer, 52, **54, 107**–08
Walter Reed Medical Center, **113,** 115
Walton, Travis, 105
War Bond Drive, 59
Washington, D.C., 68, 72
wax pieces, 52, **108**
White Sands Proving Grounds/Missile Range, 66, 68–70, 114, 119, 122–123, 128
Willow Springs, Missouri, 27
Wisconsin Ice Age, 2
*Witness to Roswell*, 125
   Winters, Chere, 135
Works Progress Administration (WPA), 60, 66
Wong, Bruce, 19
WWI gas attack, 47

## X

xylene, 53

## Z

Zamora, Lonnie, 112
Ziegelmeyer, Debbie, 18
Zigler, Charles, 134

Made in United States
Orlando, FL
18 July 2024